Springer Tracts in Natural Philosophy

Volume 11

Edited by C. Truesdell

Co-Editors: R. Aris · L. Collatz · G. Fichera · P. Germain

J. Keller · M. M. Schiffer · A. Seeger

M. M. Lavrentiev

Some Improperly Posed Problems
of Mathematical Physics

Translation revised by
Robert J. Sacker

Springer-Verlag New York Inc. 1967

Professor Dr. M. M. LAVRENTIEV
Academy of Science, Siberian Department
Novosibirsk – U.d.S.S.R.

Dr. Robert J. SACKER
New York University Courant Institute of
Mathematical Sciences

ISBN-13: 978-3-642-88212-8   e-ISBN-13: 978-3-642-88210-4
DOI: 10.1007/978-3-642-88210-4

Softcover reprint of the hardcover 1st edition 1967
Title-No. 6739

Preface

This monograph deals with the problems of mathematical physics which are improperly posed in the sense of Hadamard.

The first part covers various approaches to the formulation of improperly posed problems. These approaches are illustrated by the example of the classical improperly posed Cauchy problem for the Laplace equation.

The second part deals with a number of problems of analytic continuations of analytic and harmonic functions.

The third part is concerned with the investigation of the so-called inverse problems for differential equations in which it is required to determine a differential equation from a certain family of its solutions.

Novosibirsk June, 1967   M. M. LAVRENTIEV

Table of Contents

Chapter I

Formulation of some Improperly Posed Problems of Mathematical Physics

§ 1. Improperly Posed Problems in Metric Spaces

The notion of correctness* introduced at the beginning of our century by the French mathematician HADAMARD plays an important role in the investigation of the problems of mathematical physics. One often says that a problem is solved if its correctness is established.

Various authors present notions of correctness which coincide in their essence but differ in details. We give one of the possible definitions of correctness which is convenient for our aims.

Let Φ, F be some complete metric spaces, and let $A\varphi$ be a function with the domain of definition Φ and the range of values F. Consider the equation

$$A\varphi = f \qquad (1.1)$$

Let us point out that most problems of mathematical physics can be reduced to the investigation of the solution of equation (1.1) with a given function A and right-hand side f. We say that the problem of solving (1.1) is properly posed if the following conditions are satisfied:

1) The solution of (1.1) exists for any $f \in F$.

2) The solution of (1.1) is unique in Φ.

3) The solution of (1.1) depends continuously on the right-hand side f.

In other words, the problem of solving (1.1) is properly posed if there exists a function Bf defined and continuous over all of F, which is inverse to the function $A\varphi$.

Linear problems are most often considered in mathematical physics. In this case Φ, F are BANACH spaces, and A is a linear operator. The BANACH spaces Φ, F in concrete problems are the known functional spaces C^l, L_p, W_p^b, H_p^q, S_p, ... with the carriers in some n-dimensional space of the independent variables or on any part of the spaces of independent variables.

---

* Translator's note: Hereafter we shall refer to the *correctness* or *incorrectness* of problems which are respectively *properly posed* or *improperly posed*.

The first requirement of correctness is that the problem should not be overdetermined, and superfluous conditions should not be imposed.

The second requirement is that the solution be unique.

The third requirement of correctness, continuity of the inverse function Bf, arises from the fact that in the real problems of mathematical physics the right-hand side of equation (1.1) is obtained from measurements made with the aid of actual instruments and is therefore known only approximately. Therefore, it has been felt for a long time that if at any point f the function Bf is discontinuous, then the solution φ cannot be uniquely recovered from the right-hand side f.

HADAMARD introduced the notion of correctness by giving an example of an improperly posed problem which became classical and was included in most text-books on mathematical physics. The example is the CAUCHY problem for the LAPLACE equation. It is well-known that the solution of this linear problem does not depend continuously on the data obtained from any of the functional spaces mentioned above. On the basis of this HADAMARD concluded that CAUCHY's problem for the LAPLACE equation and, in general, all problems exhibiting a similar dependence of the solution on the right-hand side, do not correspond to any real formulations, i. e., they are not problems of mathematical physics.

It was discovered later that HADAMARD's conclusion was erroneous and many real problems of mathematical physics lead to problems which are improperly posed in the sense of HADAMARD. In particular a number of important problems of geophysics lead to the CAUCHY problem for the LAPLACE equation.

There are still further examples of linear as well as nonlinear improperly posed problems which are important in the applications, namely the solution of the heat equation for negative time and CAUCHY data on the boundary, the nonhyperbolic CAUCHY problem for the wave equation, inverse problems of potential, and a number of inverse problems for differential equations.

At present there exists a number of approaches to the investigation of improperly posed problems. We will now explain them using the above classical CAUCHY problem for the LAPLACE equation as an illustration.

We consider one of the simplest versions of the CAUCHY problem for the LAPLACE equation. Let $u(x, y)$ be a twice continuously differentiable function of the variables x, y in the rectangle

$$0 \leq x \leq \pi, \quad 0 \leq y \leq H \qquad (D)$$

satisfying the following conditions

$$\Delta u = 0, \quad (x, y) \in D \qquad (1.2)$$

$$\frac{\partial}{\partial y} u(x, 0) = u(0, y) = u(\pi, y) = 0.$$

It is required to determine its values on the segment $y = h$, $0 \le x \le \pi$ by its values on the segment $y = 0$, $0 \le x \le \pi$.

The problem stated is equivalent to the solution of equation (1.1) where φ, f stand for the functions of the variable x

$$\varphi = u(x, h)$$
$$f = u(x, 0).$$

A is a linear integral operator with the kernel

$$K_h(x, \xi) = \frac{2}{\pi} \sum_{h=1}^{\infty} \cos h^{-1} kh \cdot \sin kx \cdot \sin k\xi.$$

The functional spaces Φ, F considered in the given example are the HILBERT functional spaces with the scalar products of the form

$$(\varphi_1, \varphi_2) = \sum_0^{\infty} \lambda_k \varphi_1^k \cdot \varphi_2^k \qquad \varphi_j \in \Phi$$
$$(f_1, f_2) = \sum_0^{\infty} \mu_k f_1^k \cdot f_2^k \qquad f_j \in F \tag{1.3}$$

where φ_j^k and f_j^k are the FOURIER coefficient of the functions φ_j and f_j, and λ_k and $\mu_k > 0$ are some sequences.

We note that spaces of such form are, in particular, the spaces W_2^l with the norm

$$\|\psi\|_{W^l} = \sum_{j=0}^{l} c_j \left\| \frac{\partial^j}{\partial x^j} \psi \right\|_{L_2}.$$

It may be easily seen that the solution of (1.1) in the case under consideration is unique, but for its existence and continuity it is insufficient to let Φ and F be some W_2^l with finite l.

To make the problem properly posed one can introduce a sufficiently "strong" norm in the data space F or a sufficiently "weak" one in the space Φ. It is quite evident that the solution of (1.1) is correct in the sense of our definition, if the sequences λ_k, μ_k in (1.3) satisfy the inequality

$$\lambda_k e^{kh} \le C \mu_k \tag{1.4}$$

where C is a constant.

The inequalities (1.4), in particular, are satisfied by the following sequences λ_k, μ_k corresponding to the above possibilities

$$\lambda_k' = e^{-kh}, \quad \mu_k' = 1$$
$$\lambda_k'' = 1, \quad \mu_k'' = e^{kh}. \tag{1.5}$$

However, the above approaches have the following defects. In the first case the statement does not cover the range of problems in which the errors in the data may be considered to be small only in the "ordinary" functional spaces; in the second case solutions of the problem may be objects which are not functions in the usual sense and do not correspond to physical reality.

We now formulate some approaches to the question of correctness of problems of the type under consideration which are free of the above defects and, in our opinion, quite natural from the standpoint of applications.

The first approach consists of changing the notion of correctness, namely, to one having requirements different from 1), 2), and 3).

We now state the requirements. In addition to the spaces Φ and F and the operator A, let there be given some closed set $M \subset \Phi$.

We call the problem for the solution of (1.1) properly posed according to TYKHONOV if the following conditions are fulfilled.

1) It is a-priori known that the solution φ exists for some class of data and belongs to the given set M, $\varphi \in M$.

2) The solution is unique in a class of functions belonging to M.

3) Arbitrarily small changes of the right-hand side of f which do not carry the solution φ out of M correspond to arbitrarily small changes in the solution φ.

We denote by M_A the image of M after the application to the space Φ of the operator A.

Requirement 3) can be restated in the following manner,

3) The solution of equation (1.1) depends continuously on the right-hand side f on the set M_A.

If M is a compact set the following statement holds (see [17]).

If equation (1.1) satisfies the requirements 1), 2) of correctness due to TYKHONOV, then there exists a function $\alpha(\tau)$ such that

a) $\alpha(\tau)$ is a continuous nondecreasing function with $\alpha(0) = 0$.

b) for any $\varphi_1, \varphi_2 \in M$ satisfying the inequality

$$\varrho(A\varphi_1, A\varphi_2) \leq \varepsilon$$

the following holds

$$\varrho(\varphi_1, \varphi_2) \leq \alpha(\varepsilon).$$

Thus, the requirement of continuous dependence 3) is satisfied if 1) and 2) are satisfied.

We note that, if a problem is properly posed according to TYCHONOV and we replace the metric spaces Φ, F by their supspaces M, M_A then the problem becomes properly posed in the usual sense.

The necessity of examining spaces Φ, F together with M, M_A is due to the fact that in real problems the errors committed in the determination of the right-hand side f usually lead to f outside of M_A. The consideration of the problem according to TYKHONOV's formulation gives the possibility of constructing an approximate solution with a certain guaranteed degree of accuracy in spite of the fact that an exact solution of (1.1) with approximate data either does not exist at all or may strongly deviate from the "true" solution.

In the CAUCHY problem for the LAPLACE equation we consider as the set M the sets defined in the following manner: we denote by A_{h_1} an integral operator with the kernel

$$K_{h_1}(x, \xi) = \frac{2}{\pi} \sum_0^\infty \cos h^{-1} k h_1 \sin kx \cdot \sin k\xi$$

and let

$$\varphi \in M(l, h_1, C)$$

if

$$\varphi = A_{h_1}\psi, \quad \|\psi\|_{W_2^l} \le C, \quad (h_1 = H - h).$$

We now give evidence that our formulation is natural for the case in which the CAUCHY problem for the LAPLACE equation is used for the solution of the geophysical problem of interpreting the gravitational or magnetic anomalies. The problem of interpretation of geophysical data is as follows: the characteristics of some physical field interacting with the inner layers of the earth are measured on the earth's surface. Certain characteristics of the structure of these layers are required to be determined. In interpreting the constant magnetic and gravitational fields there often arises the following situation. An upper layer of the earth's crust consists of uniform sedimentary rocks. The thickness of the sedimentary rock layer may be estimated from below by some constant M. The anomalies observed on the earth surface, i.e., the CAUCHY data for a harmonic function, are generated by some bodies lying at a depth exceeding H and the ultimate aim of an observer is to determine these bodies. This problem is rather complicated; its solution is not unique if one proceeds only from measurements of a field on the surface. However, by proper evaluation of the general geological situation and with well chosen supplementary hypotheses, the observer often succeeds in finding solutions with satisfactory accuracy.

It is important in this problem to make good estimates on the number of the bodies, their approximate position, and their size. For the case in which the average depth of embedding of the bodies is substantially less than their horizontal dimensions and the spacing between them, the distinct extrema of the anomaly correspond to distinct bodies. However, if the average depth of imbedding roughly coincides with or exceeds the characteristics horizontal dimensions, then there is no such correspondence. It is possible for two bodies to cause one extremum situated in the gap between them. In order to obtain the case in which the depth of embedding is less than the horizontal dimensions one uses the solution of CAUCHY's problem for the LAPLACE equation. The existence of the solution is ensured by the fact that the field to be measured is generated by real objects and by the fact that in the region of evaluation, the depth of which does not exceed H, singularities of the field do not occur. The fact that the solution belongs to the set M is guaranteed by the fact that in the earth's crust one does not encounter bodies of excessive density or with magnetic intensity exceeding some completely determined constants. For example, when $u(x, y)$ is a gravitational potential the above considerations guarantee that the solution belongs to a set $M(2, h_1, C)$, in the case when u is a component of the gravitational field stress—$M(1, h_1, C)$, and in the case when u is a component of the magnetic field stress—$M(0, h_1, C)$.

This approach to the question of the formulation of improperly posed problems was originally proposed by TYKHONOV [17]. A systematic investigation of various problems, properly posed in the sense of TYKHONOV, was carried out in [28], [31], [50], [56]. A detailed description of the definitions as well as the investigation of a number of problems which are properly posed in the sense of TYCHONOV are contained in the monograph [35].

Another possible approach to the study of improperly posed problems is to let the notion of correctness remain unchanged, but change the notion of a "solution" of the problem. Besides the spaces Φ and F and the operator A let there be given, as in the preceding case, some set M. By a quasisolution of equation (1.1) we mean an element $\tilde{\varphi}$ defined by the relation

$$\varrho(A\tilde{\varphi}, f) = \min_{\varphi \in M} \varrho(A\varphi, f).$$

We note that the problem of finding a quasisolution is one of nonlinear programming. From the theorems of nonlinear programming one obtains the corresponding theorems on the correctness of the problem of finding a quasisolution.

The problems which may be regarded as properly posed according to TYKHONOV are those for which the notion of quasisolution may be suitably introduced. In some cases the quasisolution may be non-unique, but when the problem is properly posed due to TYKHONOV, any quasisolution constructed using an approximate right-hand side will be one of the best versions of an approximate solution of the problem.

The notion of a quasisolution was introduced in [45]. Even though, as previously mentioned, the range of applicability of this notion coincides with that of the notion of correctness due to TYKHONOV, the notion differs by offering greater clarity and simplicity.

The most general approach to improperly posed problems was given in the recently published works of TYKHONOV. Following [18] we define a regularizor for Eq. (1.1) to be any one-parameter family of operators B_τ having domain of definition F and range of values Φ and satisfying the following conditions

1) for any $\tau > 0$ the operator B_τ is defined and continuous over all of F.

2) for any $\varphi \in \Phi$:

$$\lim_{\tau \to 0} B_\tau A \varphi = \varphi .$$

We call the problem of solving (1.1) regularizable if there exists a regularizor for (1.1). It may be easily seen that the regularizor permits us to construct a solution with a guaranteed degree of accuracy according to the approximate right-hand side. In fact, let $\tilde{f}$ be the right-hand side of (1.1) with an error ε, i.e.,

$$\varrho(f, \tilde{f}) \leq \varepsilon .$$

We denote

$$\tilde{\varphi}_\tau = B_\tau \tilde{f}; \quad \varphi_\tau = B_\tau A \varphi$$

and estimate the quantity $\varrho(\varphi, \tilde{\varphi}_\tau)$.

By the triangle inequality

$$\varrho(\varphi, \tilde{\varphi}_\tau) \leq \varrho(\varphi, \varphi_\tau) + \varrho(\varphi_\tau, \tilde{\varphi}_\tau) . \tag{1.6}$$

It is clear that

$$\begin{aligned} \varrho(\varphi, \varphi_\tau) &\leq \alpha_1(\tau) \\ \varrho(\varphi_\tau, \tilde{\varphi}_\tau) &\leq \alpha_2(\tau, \varepsilon) \end{aligned} \tag{1.7}$$

where $\alpha_1(\tau)$ is a function characterizing the rate of the convergence of $B_\tau A \varphi$ to φ, and $\alpha_2(\tau, \varepsilon)$ is the modulus of continuity of the operator B_τ at the point $A\varphi$. The functions $\alpha_1(\tau)$ and $\alpha_2(\tau, \varepsilon)$ for fixed τ are continuous and monotonic, and $\alpha_1(0) = \alpha_2(\tau, 0) = 0$.

Substituting (1.7) into (1.6) and setting τ equal to a root of the equation

$$\alpha_1(\tau) = \alpha_2(\tau, \varepsilon) \tag{1.8}$$

where $\tilde{\alpha}_1(\tau)$, is the inverse of $\alpha_1(\tau)$, we obtain

$$\varrho(\varphi, \tilde{\varphi}_\tau) \leq 2 \alpha_1(\tau) . \tag{1.9}$$

It is evident that the root of equation (1.8), and therefore, the right-hand side in the inequality (1.9) tends to zero as $\varepsilon \to 0$.

In contrast to the first two approaches, in the notion of regularizability we consider no additional sets other than the basic spaces Φ, F. This accounts for the greater generality of this approach. Apparently, all the problems of mathematical physics connected with real phenomena are regularizable. (HADAMARD regarded all such problems to be properly posed.) In [22] the regularizability of a rather wide class of equations (1.1) has been proved.

In conclusion we point out the relation between the notion of a quasisolution and a regularizor. Suppose the solution of (1.1) is unique. Let M_τ be a one-parameter family of compact sets ($\tau > 0$) satisfying the requirements

$$M_{\tau_1} \subset M_{\tau_2}, \quad \tau_1 < \tau_2 \overline{\underset{\tau > 0}{\cup} M_\tau} = \Phi,$$

and let B_τ be a continuous operator with domain of definition F such that $B_\tau f$ is a quasisolution of (1.1) with the set $M = M_\tau$. It is evident that the operator B_τ is a regularizor for equation (1.1).

Besides the above three approaches to improperly posed problems, an approach related to the theory of probability has been developed to some extent. The next section is devoted to this approach.

§ 2. A Probability Approach to Improperly Posed Problems

The theory of probability was first utilized in the investigation of improperly posed problems in a paper by KHALFIN and SUDAKOV [66]. The authors restricted their investigation to the classical CAUCHY problem for the LAPLACE equation. The main content of the paper is as follows.

The CAUCHY data are assumed to consist of a sum of two items, the "exact data" and the "error". The "error" is a realization of a stationary random process. The question to be investigated is, under what conditions on the autocorrelation function of the error will this error insignificantly affect (in the average) the solution of a problem. It turns out to be sufficient that the spectral function of the error should be small in the HILBERT space with the exponential weight.

KHALFIN and SUDAKOV's approach is analogous to a "determinate" approach in which the class of permissible initial data (see Section 1) is narrowed. We present here a statistical approach similar to the determinate one in which the notion of correctness according to TYKHONOV or of quasi-solutions is introduced. In this case we shall use the notions and definitions of information theory presented in the paper by DOBRUSHIN [49].

As it was noted in Section 1, most of the problems of mathematical physics reduce to the examination of equation

$$A\varphi = f \tag{1.1}$$

where φ, f are elements of the metric spaces Φ, F and A is a continuous operator. In practice the right-hand side is, as a rule, a result of observations made with the aid of real physical instruments and therefore cannot be considered to be given with absolute accuracy. However, in practice it is often the case that a certain statistical distribution of errors in the definition of the right-hand side of (1.1), and the statistical distribution of solutions φ are given. Motivated by this we propose the following approach.

Let Φ, F be measurable spaces (see [49]), i.e., in the spaces Φ, F besides the metrics, there are also given some σ-algebras S_φ, S_f of subspaces of the spaces Φ, F and the operator A is measurable. In addition let Ω be a probability space, i.e., a measurable space with σ-algebra S_Ω and with probability measure $P\{\}$ given on this σ-algebra. The elements φ, f in (1.1) will be considered to be random variables with their values in the spaces Φ, F. Moreover, we consider also a random variable $\tilde{f}$ with values in F—the right-hand side of (1.1) with an error—and the random variable $(f, \tilde{f})$ with values in $F \times F$, the product of F with itself. Let B be an operator with domain of definition in F and the range of values in Φ, and let $\tilde{\varphi} = B\tilde{f}$, where φ, $\tilde{\varphi}$ are random variables in the space $\Phi \times \Phi$.

The problem of constructing an approximate solution to (1.1) consists of constructing such an operator B that the distance between φ, $\tilde{\varphi}$ be sufficiently small where the notion of smallness should take into account both the metric of the space Φ and the distribution of random variables $(\varphi, \tilde{\varphi})$.

As an estimate of the deviation of the approximate solution $\tilde{\varphi}$ from the exact solution φ one may take, for instance, the quantity $m\varrho\,(\varphi, \tilde{\varphi})^*$ or the quantity $P\{\varrho\,(\varphi, \tilde{\varphi}) > \varepsilon\}$. Corresponding to these, we introduce two notions of optimal solution of equation (1.1).

1. The function $\tilde{\varphi} = B\tilde{f}$ is called an optimal solution of (1.1) if

$$m\varrho\,(\varphi, \tilde{\varphi}) = \min_B .$$

2. The function $\tilde{\varphi} = B\tilde{f}$ is called an optimal ε solution of (1.1) if

$$P\{\varrho\,(\varphi, \tilde{\varphi}) > \varepsilon\} = \min_B .$$

If the above estimates of deviation of $\tilde{\varphi}$ from φ do not exceed the number δ we call $\tilde{\varphi}$ a solution of (1.1) with accuracy δ in the sense 1 or 2, respectively. This statement is in its content close to the decoding problem of information theory (see [49]).

We note that since in our statement the operator A is determined, the distribution of the random variable φ in Φ uniquely induces the distribution of the random variable f in F while the joint distribution of f, $\tilde{f}$ in $F \times F$ induces the joint distribution of the random variables φ, $\tilde{f}$, in the space $\Phi \times F$. Since $\tilde{f}$ in the statement is considered to be given, we may consider the dis-

---

* m is the mathematical expectation, the random variable $\varrho\,(\varphi, \tilde{\varphi})$ is assumed to be measurable.

tribution in Φ induced by a joint distribution of φ, $\tilde{f}$ at fixed $\tilde{f}$. If the distribution is to be given, the above formulated problem of constructing optimal solutions and solutions with prescribed accuracy become classical problems of the theory of probability. However, the practical construction of the joint distribution of φ, $\tilde{f}$ may involve considerable difficulties, and so in most cases it is expedient to choose simpler although less accurate methods.

The determinate approaches in which the notion of correctness according to Tykhonov or of a quasisolution are introduced, may be regarded as particular cases of the proposed approach. In fact, let the set M introduced in the previous section be measurable, and

$$P\{\varphi \in M\} = 1. \qquad (1.10)$$

Clearly the equality (1.10) may be regarded as a priori information that the solution φ exists and belongs to the set M.

Proceeding from the above ideas, we now give some results relating to optimal solutions and to solutions with a prescribed degree of accuracy, obtained under some additional assumptions.

1. Let F, Φ be Hilbert spaces. We denote by $\psi_{\tilde{f}}$ a random variable in Φ whose distribution is induced by a joint distribution of φ, $\tilde{f}$ for fixed $\tilde{f}$. One can easily see that in this case the optimal solution in the sense of 1 is given by the formula,

$$\tilde{\varphi} = \int_{\Omega} \psi_{\tilde{f}} \, d\omega \qquad (1.11)$$

In fact, for any $\tilde{\varphi}$

$$m \|\tilde{\varphi} - \psi_{\tilde{f}}\|^2 = \|\tilde{\varphi}\|^2 - 2(\tilde{\varphi}, \int_{\Omega} \psi_{\tilde{f}} \, d\omega) + \int_{\Omega} \|\psi_{\tilde{f}}\|^2 \, d\omega$$

from which follows the validity of our assertion.

2. Let the solution of (1.1) be unique, and let (1.10) hold, where M is some compact set and M_A is the image (in the space F) of the set M. Then, as noted in the previous section, the operator A^{-1} on a set M_A is uniformly continuous. Its modulus of continuity will be denoted by $\alpha(\tau)$.

Next let

$$P\{\varrho(f, \tilde{f}) > \delta\} = 0 \qquad (1.12)$$

where δ is a constant.

It can be easily seen that if ε in definition 2 of an optimal solution satisfies the inequality

$$\varepsilon \leq \alpha(\delta) \qquad (1.13)$$

then the quasisolution of (1.1) defined previously with the set M is an optimal solution in the sense of 2.

In fact, let $\tilde{\varphi}$ be a quasisolution of (1.1) with the righthand side $\tilde{f}$ and the set M.

Then, by virtue of (1.10), (1.12), (1.13) and the definitions of quasisolution, and modulus of continuity $\alpha(\tau)$, we get

$$P\{\varrho(\varphi, \tilde{\varphi}) > \varepsilon\} \leq P\{\varrho(f, \tilde{f}) > \delta\} = 0$$

which means that the quasisolution $\tilde{\varphi}$ is ε optimal in the sense of 2.

3. Let B_λ be a regularizor of (1.1).

We estimate the accuracy, in the sense of 2, of the solution of (1.1) with the aid of the regularizor B_λ. We assume, in addition, that the operator B_λ is uniformly continuous on F for any $\lambda > 0$, and denote by $\alpha(\varepsilon, \lambda)$, $\alpha_\lambda(\tau)$, $\beta(\tau)$ the functions

$$\alpha(\varepsilon, \lambda) = P\{\varrho(B_\lambda A\varphi, \varphi) > \varepsilon\}$$

$$\alpha_\lambda(\tau) = \max_{\varrho(u, v) \leq \tau} \varrho(B_\lambda u, B_\lambda v)$$

$$\beta(\tau) = P\{\varrho(\tilde{f}, f) > \tau\}$$

By virtue of the above assumptions and the known properties of the random variables

$$P\{\varrho(B_\lambda \tilde{f}, \varphi) \leq \varepsilon\} \geq \max\left[P\left\{\varrho(B_\lambda \tilde{f}, B_\lambda f) \leq \frac{\varepsilon}{2}\right\}; P\left\{\varrho(B_\lambda f, \varphi) \leq \frac{\varepsilon}{2}\right\}\right] \quad (1.14)$$

$$P\left\{\varrho(B_\lambda \tilde{f}, B_\lambda f) \leq \frac{\varepsilon}{2}\right\} \geq P\left\{\varrho(\tilde{f}, f) \leq \alpha_\lambda'\left(\frac{\varepsilon}{2}\right)\right\} = 1 - \beta\left[\alpha_\lambda'\left(\frac{\varepsilon}{2}\right)\right] \quad (1.15)$$

where $\alpha_\lambda'(\tau)$ is the function inverse to $\alpha_\lambda(\tau)$.

Substituting (1.15) in (1.13) we get

$$P\{\varrho(B_\lambda \tilde{f}, \varphi) > \varepsilon\} \leq \max\left[\beta\left[\alpha_\lambda'\left(\frac{\varepsilon}{2}\right)\right], \alpha\left(\frac{\varepsilon}{2}, \lambda\right)\right]. \quad (1.16)$$

We note that the estimate (1.16) makes possible an optimal choice of the parameter λ of the regularizor B_λ.

4. We now give the estimates of the effectiveness of a particular regularizor for the CAUCHY problem for the LAPLACE equation considered in the preceding section. Let F and Φ be the space L_2 of functions defined on the segment $[0, \pi]$ and A_h be an integral operator

$$A_h\varphi = \int_0^\pi K_h(x, \xi)\varphi(\xi)\,d\xi$$

with the kernel K_h equal to

$$K_h(x, \xi) = \frac{2}{\pi}\sum_0^\infty \cos h^{-1} kh \cdot \sin kx \cdot \sin k\xi.$$

Next, let the distribution of the random variables φ, f, $\tilde{f}$ be such that

$$P\{\|A_{h_1}^{-1}\varphi\| > \tau\} = (p^2\tau^2 + 1)^{-1}$$
$$P\{\|f - \tilde{f}\| > \tau\} = (q^2\tau^2 + 1)^{-1} \qquad (1.17)$$

As a regularizor B_λ we consider an integral operator with the kernel

$$Q_\lambda(x, \xi) = \frac{2}{\pi} \sum_0^\infty \cosh k h \, (1 + \lambda e^{kh_2})^{-1} \sin kx \cdot \sin k\xi \quad h_2 > h_1$$

One can show that the operator B_λ satisfies the following relations

$$\|B_\lambda\| \leq \lambda^{-h/h_2} \cdot r_1(h, h_2)$$
$$\|A_{h_1}^{-1}\varphi\| \geq \lambda^{-h_1/h_2} \cdot r_2(h_1, h_2) \cdot \|(E - B_\lambda A)\varphi\| \qquad (1.18)$$

where

$$r_1(h, h_2) = h^{h/h_2}(h_2 - h)^{(h_2-h)/h_2} \cdot h_2^{-1}$$
$$r_2(h_1, h_2) = (h_2 - h_1)^{-(h_2+h_1)/h_2} \cdot h_1^{-h_1/h_2} \cdot h_2$$

From (1.17) and (1.18) we get

$$P\{\|B_\lambda(\tilde{f} - f)\| > \varepsilon\} \leq P\{\|B_\lambda\| \cdot \|\tilde{f} - f\| > \varepsilon\} = (q^2\lambda^{2h/h_2} \cdot \varepsilon^2 r_1^{-2} + 1)^{-1},$$
$$P\{\|B_\lambda A - E)\varphi\| > \varepsilon\} \leq P\{\|A_{h_1}^{-1}\varphi\| > \lambda^{-h_1/h_2} \cdot r_2 \cdot \varepsilon\} = (p^2\lambda^{-2h_1/h_2} \cdot r_2^2 + 1)^{-1},$$

from which it follows that

$$P\{\|B_\lambda\tilde{f} - \varphi\| > 2\varepsilon\} \leq \max\left[(q^2\lambda^{2h/h_2} \cdot \varepsilon^2 r_1^{-2} + 1)^{-1}; (p^2\lambda^{-2h_1/h_2} \cdot r_2^2 + 1)^{-1}\right]. \qquad (1.19)$$

It is evident that the first expression in the right-hand side of (1.19) monotonically decreases as λ increases, while the second one monotonically increases. This means that in the sense of the estimate (1.19), the value of the parameter of the regularizor will be optimal when the two expressions coincide. By equating the expressions in the right-hand side of (1.19) we obtain

$$\lambda_0 = \left(\frac{p}{q} \cdot \frac{r_1 r_2}{\varepsilon}\right)^{h_2/(h+h_1)}$$

$$P\{\|B_{\lambda_0}\tilde{f} - \varphi\| > 2\varepsilon\} \leq (p^{h/(h+h_1)} \cdot q^{h_1/(h+h_1)} \cdot \varepsilon^{h_1/(h+h_1)} \cdot r_3 + 1)^{-1}. \qquad (1.20)$$

From (1.20) it follows that for any δ, $\varepsilon > 0$ there can be found a sufficiently large q—the index of the "quality of the measurements"—so that the estimated probability will be less than δ.

We also note that the regularizor B_λ used by us also permits optimality with respect to the parameter h_2, but the analytical determination of the optimal h_2 leads to extremely complex calculations. It seems that the parameter should be chosen on the interval $[H, 2H]$.

Chapter II

Analytic Continuation

The problem of analytic continuation is investigated in this chapter. Estimates of "stability" for the solution of some problems are given, and concrete solutions are constructed.

§ 1. Analytic Continuation of a Function of One Complex Variable from a Part of the Boundary of the Region of Regularity

The problem was first considered by CARLEMAN in [3]. The results were obtained in [5].

I. The First Problem.

Let $f(z)$ be an analytic function, regular and bounded in some bounded domain D so that

$$|f(z)| \leq M \quad z \in D \tag{2.1}$$

Γ is the boundary of D; Γ' is a part of Γ and $\Gamma'' = \Gamma - \Gamma'$. Let the values of $f(z)$ on Γ' be known, and suppose it is required to determine $f(z)$ in some part of D. We now prove a theorem characterizing the stability of the solution of the problem.

Theorem 1. Let $f(z)$ on the curve Γ' satisfy the inequality

$$|f(z)| \leq \varepsilon \quad (z \in \Gamma') \tag{2.2}$$

Then the inequality

$$|f(z)| \leq M^{1-\omega(z)} \cdot \varepsilon^{\omega(z)} \tag{2.3}$$

will also hold where $\omega(z)$ is the harmonic measure of the curve Γ' with respect to the point z and the domain D.

For the proof we consider the function

$$\varphi(z) = \ln|f(z)|.$$

It is known that $\varphi(z)$ is a harmonic function. From (2.1), (2.2) it follows that the function $\varphi(z)$ satisfies the inequalities

$$\varphi(z) \le \ln \varepsilon \qquad z \in \Gamma'$$
$$\varphi(z) \le \ln M \qquad z \in \Gamma''$$

(2.4)

from which it follows that

$$\varphi(z) \le \omega(z) \ln \varepsilon + [1 - \omega(z)] \ln M$$ (2.5)

The inequality (2.3) to be proved follows from (2.5).

The problem we are considering can be reduced to an integral equation of the first kind and can be solved by one of the general methods. In fact, we denote by D' the part of D in which $f(z)$ is to be determined. We denote by P, Q the ends of the curve Γ'' and consider some points P_j, Q_j ($j = 1, 2$) lying on Γ'' sufficiently close to the points P, Q, respectively. We denote by Γ_j'' the part of Γ'' lying between the points P_j, Q_j, and let the location of the points P_j, Q_j be such that $\Gamma_2'' \in \Gamma_1''$ (see Figure 1).

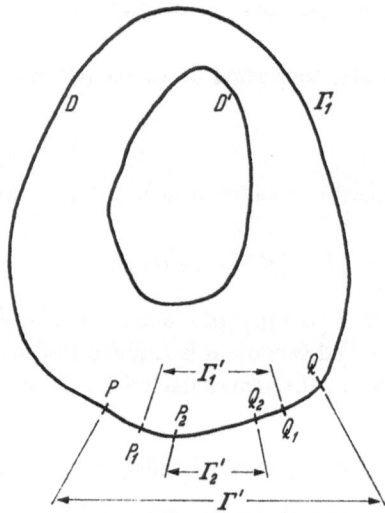

We draw a curve Γ_1 through P_1, Q_1 which lies inside D so that the domain D' lies inside the domain D_1' bounded by the curves Γ_1, Γ_1''; $D' \in D_1'$. It is evident that in order to determine the function $f(z)$ for any $z \in D'$ it is sufficient to determine $f(z)$ on the curve Γ_1 in view of the CAUCHY formula at $z \in D'$

$$f(z) = \frac{1}{2\pi i} \int_{\Gamma_1 \cup \Gamma_1'} \frac{f(\zeta)}{\zeta - z} d\zeta$$

and the $f(z)$ values given on Γ_1''.

We denote by $f_2(z)$ the limiting values on the curve Γ_2' of the CAUCHY-like integral of the function $f(z)$ on the arc Γ'' as z tends to Γ_2' from inside of D;

and by $f_1(z)$ we denote the values on the curve Γ_1 of the CAUCHY-like integral of the function $f(z)$ on the arc Γ''

$$f_2(z)=\frac{1}{2\pi i}\lim_{\substack{z\to\Gamma_2'\\z\in D}}\int_{\Gamma'}\frac{f(\zeta)}{\zeta-z}d\zeta;\ f_1(z)=\frac{1}{2\pi i}\int_{\Gamma'}\frac{f(\zeta)}{\zeta-z}d\zeta,\quad z\in\Gamma_1.$$

The limiting values of the CAUCHY-like integral are calculated by the familiar SOKHOTSKY formulas, so that for an investigation of this problem we may suppose that the functions f_2, $\tilde{f}_1$ are given.

We denote by A the operator transforming the arbitrary complex function $\varphi(z)$ on Γ_1 to the function $\psi(z)$ given on Γ_1' by the formula

$$\psi(z)=\frac{1}{2\pi i}\int_{\Gamma_1}\frac{\varphi(\zeta)}{\zeta-z}d\zeta,\quad z\in\Gamma_1'.$$

It can be easily seen that the validity of the equation

$$A\varphi(x)=f(z)-f_2(z)\quad z\in\Gamma_2'$$

is necessary and sufficient for the value of the function $\varphi(z)$ to coincide with the values on Γ_1 of the unknown function $f(z)$. Thus, the problem of determining the function $f(z)$ on Γ_1 is equivalent to that of solving the linear operator equation of the first kind

$$A\varphi=f-f_2.$$

Then by the CAUCHY formula for any $z\in\Gamma_1$

$$f(z)=\frac{1}{2\pi i}\int_{\Gamma''}\frac{f(\zeta)}{\zeta-z}d\zeta+f_1(z)$$

from which it follows that the function $f(z)$ on Γ_1 is equal to

$$f(z)=BW(z)+f_1(z),$$

where $W(z)$ is a function defined on Γ''

$$\|W(z)\|\leq M,$$

(M is a constant depending on D), and B is the operator transforming the complex function $W(z)$ given in Γ'' to the function $BW(z)$ given on Γ_1 by the formula

$$BW(z)=\frac{1}{2\pi i}\int_{\Gamma''}\frac{W(\zeta)}{\zeta-z}d\zeta.$$

It is clear that B is a linear completely continuous operator and therefore in the problem of solving the equation

$$A\varphi = f - f_2$$

the conditions under which one treats linear operator equations of the first kind are met.

We now consider another method of concrete solution based on the integral formulas of CARLEMAN type. We call a CARLEMAN function for the domain D and the curve Γ' any function $G(z, \zeta, \delta)$ having the following properties

1) $$G(z,\zeta,\delta) = \frac{1}{\zeta - z} + \tilde{G}(z,\zeta,\delta)$$

where $\tilde{G}(z, \zeta, \delta)$ is an analytic function of the variable ζ, regular and bounded in the domain D.

2) the function $G(z, \zeta, \delta)$ satisfies the inequality

$$\int_{\Gamma''} |G(z,\zeta,\delta)| \, |d\zeta| \le \delta.$$

We will construct the concrete solution for the problem with the aid of a CARLEMAN function. Let the values of $f(z)$ on the curve Γ' be known with accuracy ε, i.e., a function $f_\varepsilon(z)$ is known on the curve Γ' such that

$$|f_\varepsilon(z) - f(z)| \le \varepsilon, \quad z \in \Gamma'.$$

We denote by the function $f_\delta(z)$

$$f_\delta(z) = \frac{1}{2\pi i} \int_{\Gamma'} G(z,\zeta,\delta) f_\varepsilon(\zeta) \, d\zeta,$$

and estimate the difference $f(z) - f_\delta(z)$.

According to CAUCHY's formula

$$f(z) = \frac{1}{2\pi i} \int_{\Gamma} G(z,\zeta,\delta) f(\zeta) \, d\zeta.$$

Therefore

$$f(z) - f_\delta(z) = \frac{1}{2\pi i} \int_{\Gamma''} G(z,\zeta,\delta) f(\zeta) \, d\zeta$$

$$+ \frac{1}{2\pi i} \int_{\Gamma'} G(z,\zeta,\delta) [f(\zeta) - f_\varepsilon(\zeta)] \, d\zeta. \tag{2.6}$$

In virtue of (2.1) and the second property of CARLEMAN functions the first integral in the right-hand side of (2.6) satisfies the inequality

$$\left| \int_{\Gamma''} Gf(\zeta) \, d\zeta \right| \le \delta \cdot M. \tag{2.7}$$

In virtue of (2.2) the second integral in the right-hand side of (2.6) satisfies the inequality

$$|\int_{\Gamma'} G[f(\zeta)-f_\varepsilon(\zeta)]\,d\zeta|\leq\mu(z,\delta)\cdot\varepsilon\cdot\gamma',\qquad(2.8)$$

where γ' is the length of the curve Γ',

$$\mu(z,\delta)=\max_{\zeta\in D}|G|.$$

Substituting (2.7), (2.8) in (2.6) we obtain

$$|f(z)-f_\delta(z)|\leq\frac{1}{2\pi}[\delta M+\mu(z,\delta)\cdot\varepsilon\cdot\gamma'].\qquad(2.9)$$

Let the CARLEMAN function be known. We denote the root of the equation

$$\frac{\delta}{\mu(z,\delta)}=\frac{\varepsilon\gamma'}{M}\qquad(2.10)$$

by $\delta(z,\varepsilon)$, and put

$$\delta=\delta(z,\varepsilon)$$

in (2.9). Then the inequality (2.9) becomes

$$|f(z)-f_\delta(z)|\leq\frac{1}{\pi}M\cdot\delta(z,\varepsilon)\qquad(2.11)$$

It is evident that as $\varepsilon\to 0$, $\delta(z,\varepsilon)$ and therefore the right-hand side of (2.11) tends to zero. Hence, if the CARLEMAN function $G(z,\zeta,\delta)$ is known, the function $f_{\delta\varepsilon}(z)$ may be considered to be an approximate solution of the problem. The difference between the "true" solution $f(z)$ and the approximate one $f_{\delta\varepsilon}(z)$ is thus estimated by the above inequality (2.11).

It is evident that the smaller the function $\mu(z,\delta)$ is, the higher will be the precision of the approximate solution obtained by the above method. We shall now display a family of CARLEMAN functions with minimal $\mu(z,\delta)$ in the case where D is a simply connected domain.

Let $\theta_1(z,\zeta)$ be the function which is harmonic with respect to the variable ζ, regular in the domain D and on Γ takes the boundary values

$$\theta_1(z,\zeta)=\ln|z-\zeta|,\quad\zeta\in\Gamma.$$

We denote by $\theta_2(z,\zeta)$ the harmonic function conjugate with respect to $\theta_1(z,\zeta)$ and by $\theta(z,\zeta)$ we denote the analytic function

$$\theta(z,\zeta)=\theta_1(z,\zeta)+i\theta_2(z,\zeta)-\theta_1(z,z)-i\theta_2(z,z).$$

Let $\omega(z)$ the harmonic function conjugate with respect to the harmonic measure $\bar\omega(z)$, and $W(z)$ the analytic function

$$W(z)=\omega(z)+i\bar\omega(z).$$

We consider now the function

$$G(z,\zeta,\delta)=\frac{1}{\zeta-z}\exp\left\{\frac{\theta_1(z,z)+\ln\delta}{1-\omega(z)}W(\zeta)+\theta(z,\zeta)-q(z,\delta)\right\}$$

where

$$q(z,\delta)=\frac{\theta_1(z,z)+\ln\delta\cdot\omega(z)}{1-\omega(z)}+i\left[\frac{\theta_1(z,z)+\ln\delta}{1-\omega(z)}\overline{\omega}(z)+\theta_2(z,z)\right].$$

It can be easily seen that this function is a CARLEMAN function,

$$\mu(z,\delta)=\exp\left\{-\frac{\theta_1(z,z)+\ln\delta\cdot\omega(z)}{1-\omega(z)}\right\}. \tag{2.12}$$

The constructed CARLEMAN function is best possible from the standpoint of estimating the precision of the approximate solution by (2.11). However, it turns out that for certain domains the construction of the function is somewhat cumbersome. Therefore, in order to solve the problem it is preferable, in most cases, to use another CARLEMAN function, namely ,the one pointed out by CARLEMAN himself (see. ref. [3], [29]). This function does not give high accuracy, but its actual construction is rather simple.

§ 2. The Cauchy Problem for the Laplace Equation

The CAUCHY problem for the LAPLACE equation in the plane is similar to that of analytic continuation considered in § 1. An analytic function is to be determined from its values on a curve on which CAUCHY data are given.

In fact, let the value of the harmonic function $u(x,y)$ and its normal derivative $\partial/\partial n\, u(x,y)$ be known on some curve Γ. We denote by $f(z)$ ($z = x + iy$) the function $f(z) = u + iv$ where v is the function conjugate to u. It is well known that on the curve Γ

$$v(z)=\int_{z_0}^{z}\frac{\partial}{\partial n}u(z)\,ds+C$$

where z_0 is one of the endpoints of Γ.

Hence, if $u(z)$, $\partial/\partial n\, u(z)$ are known on Γ, one may consider that the values of the analytic function $f(z)$ on Γ are known.

This section deals with the extension of the CARLEMAN function method for the case of harmonic functions in a three-dimensional space.

Let D be some domain of three-dimensional space bounded by a sufficiently smooth and closed surface Σ. Let Σ' be a part of Σ, and

$$\Sigma''\cup\Sigma'=\Sigma.$$

We call a CARLEMAN function for the domain D and the surface Σ any function of the three-dimensional vectors x, ξ and the scalar parameter δ satisfying the following conditions

1) $$G(x, \xi, \delta) = \frac{1}{r(x, \xi)} + G_1(x, \xi, \delta)$$

where $r(x, \xi)$ is the length of the vector $x - \xi$, and $G_1(x, \xi, \delta)$ is a harmonic function of ξ, regular and bounded together with its gradient in the domain D.

2) The function $G(x, \xi, \delta)$ satisfies the inequality

$$\frac{1}{4\pi} \int_{\Sigma''} \left\{ |G| + \left| \frac{\partial}{\partial n_\xi} G \right| \right\} d\sigma_\xi \le \delta \qquad (2.13)$$

where $\partial / \partial n_\xi$ is the normal derivative with respect to the surface Σ, and $d\sigma_\xi$ is the element of area on Σ with respect to the variable ξ.

Let the CARLEMAN function $G(x, \xi, \delta)$ for the domain D and the surface Σ be known. By making use of the function $G(x, \xi, \delta)$ we can prove a "stability" theorem and carry out the method of concrete solutions of the CAUCHY problem for the LAPLACE equation in three-dimensional space. We denote by $\mu(x, \delta)$ the function

$$\mu(x, \delta) = \max_{\xi \in \Sigma'} \left[|G| + \left| \frac{\partial}{\partial n_\xi} G \right| \right].$$

Theorem. Let the harmonic function $u(x)$ be regular inside D and continuously differentiable over the closure of D, and let it satisfy the inequalities

$$|u(x)| + \left| \frac{\partial}{\partial n_x} u(x) \right| \le \varepsilon \qquad x \in \Sigma'$$

$$|u(x)| + \left| \frac{\partial}{\partial n_x} u(x) \right| \le M \qquad x \in \Sigma'' \qquad (2.14)$$

Then the inequality

$$|u(x)| \le 2 M \tau(x, \varepsilon) \qquad (2.15)$$

also holds; where $\tau(x, \varepsilon)$ is a root of the equation

$$\mu(x, \tau) \cdot \varepsilon = M \tau \qquad (2.16)$$

It is evident that $\tau(x, \varepsilon) \to 0$ as $\varepsilon \to 0$.

It is clear that the value of the harmonic function $u(x)$ at a point x inside D is expressible in terms of the u, $\partial / \partial n\, u$ values on the boundary D according to GREEN's formula

$$u(x) = \frac{1}{4\pi} \int_{\Sigma'} \left\{ G \frac{\partial}{\partial n} u - \frac{\partial}{\partial n} G \cdot u \right\} d\sigma_\xi + \frac{1}{4\pi} \int_{\Sigma''} \left\{ G \cdot \frac{\partial}{\partial n} u - \frac{\partial}{\partial n} G \cdot u \right\} d\sigma_\xi$$

from which, by virtue of (2.13) and (2.14) it follows that

$$|u(x)| \leq \mu(x,\delta) \cdot \varepsilon + M \cdot \delta. \tag{2.17}$$

Assuming that δ in (2.17) is equal to a root of equation (2.16) we obtain inequality (2.15).

The construction of the concrete solution with the aid of the CARLEMAN function is completely similar to that given in Section 1. Let the CAUCHY data for the function $w(x)$ on the surface Σ' be known with accuracy ε, i.e., the functions $\varphi_\varepsilon(x)$, $\psi_\varepsilon(x)$ are known on the surface Σ' so that

$$|\varphi_\varepsilon(x) - u(x)| \leq \varepsilon,$$

$$x \in \Sigma' \tag{2.18}$$

$$\left| \psi_\varepsilon(x) - \frac{\partial}{\partial n} u(x) \right| \leq \varepsilon,$$

and let the function $u(x)$ on the surface Σ''' be known to be bounded together with its normal derivative so that

$$|u(x)| + \left| \frac{\partial}{\partial n} u(x) \right| \leq M, \qquad x \in \Sigma''. \tag{2.19}$$

We consider the function

$$u_\delta(x) = \frac{1}{4\pi} \int\limits_{\Sigma'} \left[G \cdot \psi_\varepsilon(\xi) - \frac{\partial}{\partial n} G \cdot \varphi_\varepsilon(\xi) \right] d\sigma_\xi,$$

and estimate the difference $u(x) - u_\delta(x)$. Substituting the corresponding integral expressions for u, u_δ and utilizing inequalities (2.13), (2.18), and (2.19), we have

$$|u(x) - u_\delta(x)| = \left| \frac{1}{4\pi} \int\limits_{\Sigma''} \left[G \cdot \frac{\partial}{\partial n} u - \frac{\partial}{\partial n} G \cdot u \right] d\sigma_\xi \right.$$

$$+ \frac{1}{4\pi} \int\limits_{\Sigma'} \left\{ G \left[\frac{\partial}{\partial n} u - \psi_\varepsilon \right] - \frac{\partial}{\partial n} G \left[u - \varphi_\varepsilon \right] \right\} d\sigma_\xi \Bigg| \tag{2.20}$$

$$\leq 2 M\delta + \mu(x,\delta) \varepsilon \cdot \frac{S'}{4\pi}$$

where S' is the area of the surface Σ'.

Let δ be a root of equation (2.16). Then inequality (2.20) takes the form

$$|u(x) - u_\delta(x)| \leq M\tau(x,\varepsilon) \left(2 + \frac{S'}{4\pi} \right). \tag{2.21}$$

Hence, if the CARLEMAN function G is known, the function $u_\delta(x)$ will be an approximate solution with accuracy estimated by the inequality (2.21).

We now turn to the problem of constructing the CARLEMAN function. The existence of the CARLEMAN function for an arbitrary simply connected domain D and for any part of the boundary Σ' bounded by a smooth curve follows immediately from results of MERGELYAN, [47]. In that work it was established that for any arbitrary pair of continuous functions $\varphi(x)$, $\psi(x)$ defined on a smooth piece of the surface Σ' and for any $\delta > 0$ there exists a harmonic polynomial $H_m(x)$ of order m satisfying the inequalities

$$|H_m(x) - \varphi(x)| \leq \delta$$
$$\left| \frac{\partial}{\partial n} H_m(x) - \psi(x) \right| \leq \delta. \qquad (2.22)$$

In this paper an algorithm is given for the construction of the polynomial.

We consider the function $[r(x, \xi)]^{-1}$, and let the point x lie inside the domain D. We fix δ and construct a harmonic polynomial $P_m(x, \xi)$ of order m with respect to the variables ξ_k satisfying the inequalities

$$|r^{-1}(x, \xi) - P_m(x, \xi)| \leq \delta$$
$$\xi \in \Sigma''$$
$$\left| \frac{\partial}{\partial n} r^{-1}(x, \xi) - \frac{\partial}{\partial n} P_m(x, \xi) \right| \leq \delta.$$

We denote by $G(x, \xi, \delta)$ the function

$$G(x, \xi, \delta) = r^{-1}(x, \xi) - P_m(x, \xi). \qquad (2.23)$$

It is evident that the function G defined by equality (2.23) is the CARLEMAN function for domain D and surface Σ'.

The above method for the construction of CARLEMAN functions is distinguished by its great generality. However, the algorithm for such a construction is very complicated. We now give a simpler method for the construction of a CARLEMAN function for certain values of x.

We denote by $\Omega(x, \eta)$ the sphere with center at the point η and intersecting the point x. Let the point $x \in D$ satisfy the condition: There exists a point η such that the surface Σ'' lies inside the sphere $\Omega(x, \eta)$. By expanding the function $r^{-1}(x, \xi)$ in TAYLOR series in the variables $\xi_k - \eta_k$, we obtain

$$r^{-1}(x, \xi) = \sum_0^\infty P_k(x, \eta, \xi - \eta) \qquad (2.24)$$

where P_k is a homogeneous harmonic polynomial of order k in the variables $\xi_j - \eta_j$ $(j = 1, 2, 3)$. We denote by G_n the function

$$G_n(x, \eta, \xi) = r^{-1}(x, \xi) - \sum_0^n P_k(x, \eta, \xi - \eta).$$

It may be easily seen that the function G_n satisfies the inequality

$$|G_n|+|\text{grad } G_n| \le \frac{2n|\xi-\eta|^n}{|x-\eta|^{n+1}(|x-\eta|-|\xi-\eta|)^2}. \tag{2.25}$$

In fact, on any ray $\xi - \eta = 2\tau y$ where τ is a scalar and y is a unit vector, the sum

$$\sum_0^n P_k(x,\eta,\xi-\eta) = \sum_0^n P_k(x,\eta,\tau,y)$$

is a partial sum of the TAYLOR series in the variable τ for the analytic function

$$F(\tau,x,\eta,y) = \frac{1}{\sqrt{|x-\eta|^2+2\tau(x-\eta,y)+\tau^2}}.$$

The inequality (2.25) then follows from well know estimates for the reminder term for elementary analytic functions.

From (2.25) it follows that if the number n satisfies the inequality

$$S'' \frac{2nR^n(\Sigma'')}{|x-\eta|^{n+1}} \cdot \frac{1}{[|x-\eta|-R(\Sigma'')]^2} \le \delta \tag{2.26}$$

where S'' is the area of the surface Σ'' and

$$R(\Sigma'') = \max|\xi-\eta|, \qquad \xi \in \Sigma'', \tag{2.27}$$

then the function G_n is the CARLEMAN function for the domain D and the surface Σ'.

§ 3. Determination of an Analytic Function from its Values on a Set Inside the Domain of Regularity

Let $f(z)$ be an analytic function, regular in the unit disc D,

$$|f(z)| \le 1, \qquad z \in D \tag{2.28}$$

and let A be some set inside the disc

$$|z| \le R < 1 \qquad (D_R).$$

The problem is to determine $f(z)$ in the entire disc D_R from its values on A. It is evident that for the solution of our problem it is sufficient to determine $f(z)$ on the circumference

$$|z| = R \qquad (\Gamma_R)$$

since according to the CAUCHY theorem

$$f(z) = \frac{1}{2\pi i} \int_{\Gamma_R} \frac{1}{\zeta - z} f(\zeta) \, d\zeta \,. \tag{2.29}$$

We denote by L the operator which transforms the function given on Γ_R to the function given on the set A by the formula (2.29), and by L_1 the operator which transforms, by means of the formula

$$f(z) = \frac{1}{2\pi i} \int_{\Gamma_1} \frac{1}{\zeta - z} f(\zeta) \, d\zeta \,,$$

the function given on Γ_1 to the function given on Γ_R. It is evident that L, L_1 are linear, completely continuous operators.

Solving this problem is equivalent to solving

$$L\varphi = f \tag{2.30}$$

where f is the known function defined on A, and φ is the function to be found on Γ_R, and

$$\varphi = L_1 \psi \tag{2.31}$$

where ψ is a function defined on Γ_1 satisfying the inequality

$$|\psi| \leq 1 \,. \tag{2.32}$$

Hence, the solution of the problem is reduced to the solution of an integral equation of the first kind (2.30), i.e., to the determination of the function φ from the known function f, provided it is known that the function φ satisfies the conditions (2.31), (2.32).

Now, we prove a theorem which characterizes the stability of the solution of the problem.

We introduce first some definition sand notation. Let A be a set of points on some curve. Let us denote by $H_n(A)$ a system of n intervals containing the set A

$$H_n(A) \supset A$$

and by $\mu(H_n)$ the linear measure of H_n. We call the infimum of $\mu(H_n)$ for all possible H_n containing A the n-measure of the set A and denote the n-measure of A by

$$\mu_n(A) = \inf \mu(H_n(A)) \,.$$

It should be noted that this n-measure is unusual since, in general, it does not possess the property of additivity. However, it is evident that n-measure possesses the following property which replaces in some sense the additivity

$$\mu_{n_1}(A_1) + \mu_{n_2}(A_2) \geq \mu_{n_1 + n_2}(A_1 + A_2) \,.$$

Now, let A be a set of points of the complex plane lying in the disc D_R. We denote by A_r the radial projection of the set A onto the circumference

$$|z|=r.$$

Let $W(z)$ be a conformal mapping carrying the unit disc onto itself, and carrying the set A onto the set $A_w \subset D_{r\varrho}$ where $D_{r\varrho}$ is the annulus

$$0 < r \le |z| \le \varrho < 1 \qquad (D_{r\varrho})$$

By $\mu_n^{r\varrho}(A)$ we denote the supremum of $\mu_n(A_{wr})$ taken over all possible mappings $W(z)$

$$\mu_n^{r\varrho}(A) = \sup \mu_n(A_{wr}).$$

Theorem. Let A be a set lying inside the disc

$$|z| \le R \le 1 \qquad (D_R)$$

and let the analytic function $f(z)$, regular in D, satisfy the inequalities

$$|f(z)| \le 1 \qquad (z \in D)$$
$$|f(z)| \le \varepsilon \qquad (z \in A)$$

Then the inequality

$$|f(z)| \le \exp\left\{ -C_1 \frac{\ln \varepsilon (1-|z|)}{\ln \mu_p^{r\varrho}(A)} \right\} \qquad (2.33)$$

is valid, where C_1 is some constant dependent of the radii ϱ, r, R and p satisfies the inequalities

$$\left[\frac{\mu_p^{r\varrho}(A)}{C_2} \right]^p \le \varepsilon \le \left[\frac{\mu_{p-1}^{r\varrho}(A)}{C_2} \right]^{p-1} \qquad (2.34)$$

with C_2 also dependent on ϱ, r, R.

First we prove two lemmas.

Lemma 1. For any points z, a lying in the annulus $D_{r\varrho}$ the following inequality is valid

$$\left| \frac{a-z}{1-\bar{a}z} \right| \ge \left| \frac{a_r - z_r}{1 - \bar{a}_r z_r} \right|$$

where a_r, z_r are the radial projections of the points a, z on the circle Γ_r

$$|z|=r.$$

Without loss of generality we may consider the number a to be real and positive.

Let first $|z| > a$. We consider the function

$$I_1(\tau) = \left| \frac{a - \tau z}{1 - a\tau z} \right|^2.$$

Differentiating the function $I_1(\tau)$ we obtain

$$I_1'(1)=2\frac{(1-a^2)[(x^2+y^2)(1+a^2-ax)-ax]}{|1-\bar{a}\cdot z|^4}\qquad(z=x+iy)$$

Let

$$x\leq a$$

then

$$(x^2+y^2)(1+a^2-ax)-ax>x^2+y^2-a^2>0.$$

Now let

$$x>a$$

then

$$(x^2+y^2)(1+a^2-ax)-ax>(x-a)(1-ax)x>0.$$

Hence

$$I_1'(1)>0$$

from which it follows that for $|z|>a$,

$$\left|\frac{a-z}{1-\bar{a}z}\right|>\left|\frac{a-z_{|a|}}{1-\bar{a}z_{|a|}}\right|.\tag{2.35}$$

Now, let

$$|z|=|a|.$$

We consider the function

$$I_2(\tau)=\tau\left|\frac{a-z}{1-\tau\bar{a}z}\right|^2.$$

Differentiating the function $I_2(\tau)$ we obtain

$$I_2'(1)=\frac{[(x-a)^2+y^2](1-a^4)}{(1-a^2)^4}>0$$

from which it follows that for $|z|=|a|$

$$\left|\frac{a-z}{1-\bar{a}z}\right|\geq\left|\frac{a_r-z_r}{1-\bar{a}_rz_r}\right|.\tag{2.36}$$

Finally, for $|z|<|a|$, we consider the function

$$I_3(\tau)=\left|\frac{\tau-z}{1-\tau z}\right|^2.$$

Differentiating the function $I_3(\tau)$ we have

$$I_3'=2\frac{[a^2(x^2+y^2)-2ax+1](a-x)-(x^2+y^2-2ax+a^2)(a\cdot|z|^2-x)}{|1-\bar{a}z|^4}.$$

Let

$$a(x^2+y^2)-x\le 0$$

then, clearly

$$I_3'(a)>0.$$

Now, let

$$a(x^2+y^2)-x>0$$

then

$$a-x>a(x^2+y^2)-x$$

and, hence

$$I_3'(a)>2\frac{[a(x^2+y^2)-x](1-a^2)[1-(x^2+y^2)]}{|1-\bar{a}z^4|}>0$$

from which it follows that for $|a|>|z|$

$$\left|\frac{a-z}{1-\bar{a}z}\right|>\left|\frac{a_{|z|}-z}{1-\bar{a}_{|z|}\cdot z}\right|. \qquad (2.37)$$

The inequality to be proved follows from inequalities (2.35), (2.36), and (2.37).

Lemma 2. Let a_1, a_2, z_1, z_2 be some complex numbers, and

$$|a_1|=|a_2|=|z_1|=|z_2|=r<1.$$

Then, if the inequality

$$\arg\frac{a_1}{z_1}\le\arg\frac{a_2}{z_2}\le\pi$$

is true, then the inequality

$$\left|\frac{z_1-a_1}{1-\bar{a}_1z_1}\right|\le\left|\frac{z_2-a_2}{1-\bar{a}_2z_2}\right|$$

will also be true. In proving the Lemma, one may suppose that

$$a_1=a_2=a$$

where a is real and positive.

Let

$$z_1=ae^{i\varphi_1};\quad z_2=ae^{i\varphi_2}.$$

We consider the function

$$I(\varphi)=\left|\frac{z-a}{1-\bar{a}z}\right|^2.\quad z=ae^{i\varphi}$$

Differentiating the function $I(\varphi)$ we obtain

$$I'(\varphi) = \frac{a^2 \sin \varphi \, (1-a^2)^2}{|1-\bar{a}z|^4} \geq 0$$

from which we obtain the inequality to be proved.

We now prove the theorem. Let $w = w(z)$ be the conformal mapping of D onto itself for which

$$\mu_p^{r\varrho}(A) = \mu_p(A_{wr})$$

Under this mapping, the set A is carried into the set

$$A_w \subset D_{r\varrho}.$$

Let us consider the function

$$\varphi(z) = \ln|f(z)|. \tag{2.38}$$

It is known that $\varphi(z)$ is a harmonic function, regular at those points of D where the function $f(z)$ is not zero. We consider some point a belonging to A_w.

According to the condition of the theorem

$$|f(a)| = \varepsilon_a \leq \varepsilon \tag{2.39}$$

From (2.38), (2.39) and well known properties of harmonic functions it follows that

$$|f(z)| = \varepsilon_a$$

on some analytic curve γ_a intersecting the point a (if $f(a) = 0$ then the curve consists of the single point a).

Let the curve γ_a be an open arc in the annulus

$$\frac{2}{3} r \leq |z| \leq \frac{1}{3}(1+2\varrho) \qquad \tilde{D}_{r\varrho}$$

Then the curve γ_a either intersect one of the two circles

$$|z| = \frac{2}{3} r \qquad (\tilde{\Gamma}_r)$$

$$|z| = \frac{1}{3}(1+2\varrho) \qquad (\tilde{\Gamma}_\varrho)$$

or bounds a domain containing $D_{2/3\,r}$.

Suppose γ_a intersects the circle $\tilde{\Gamma}_r$ in some point a_1. The distance between the points a, a_1 satisfies the inequality

$$|a - a_1| \geq \frac{1}{3} r. \tag{2.40}$$

It is not difficult to show that by virtue of (2.4) the harmonic measure of the curve γ_a in the domain D with respect to the point z satisfies the inequality

$$\omega(z,\gamma_a) \geq \frac{r}{12\pi} \frac{1-|z|}{1+|z|}$$

from which it follows that

$$\varphi(z) \leq \frac{r}{12\pi} \frac{1-|z|}{1+|z|} \ln \varepsilon \tag{2.41}$$

From (2.41) we obtain

$$|f(z)| \leq \exp\left\{\ln \varepsilon \frac{r}{12\pi} \frac{1-|z|}{1+|z|}\right\}. \tag{2.42}$$

In case the curve γ_a intersects the ring

$$|z| = \frac{1}{3}(1+2\varrho)$$

we similarly obtain the inequality

$$|f(z)| \leq \exp\left\{\ln \varepsilon \frac{1-\varrho}{12\pi} \frac{1-|z|}{1+|z|}\right\}. \tag{2.43}$$

Finally, if the curve γ_a bounds a domain containing $D_{2/3\,r}$ then from the maximum modulus principle, in the whole domain $D_{2/3\,r}$ the inequality

$$|f(z)| \leq \varepsilon \qquad z \in D_{\frac{2}{3}r}$$

is valid, from which the assertion of the theorem immediately follows.

Thus, if just one of the curves γ_a is not closed in the annulus $\tilde{D}_{r\varrho}$, the statement of the theorem follows. Now suppose all the curves γ_a are closed in $\tilde{D}_{r\varrho}$ and consider any curve γ_a. As γ_a is closed in D, it follows that inside the domain bounded by the curve γ_a there is at least one zero of the function $f(z)$. We denote by γ the sum of sets γ_a, and by γ_r the projection of γ on the circle

$$|z| = \frac{1}{3}r.$$

In the ring

$$\frac{1}{3}r \leq |z| \leq \frac{1}{3}(2+\varrho) \qquad \bar{D}_{r\varrho}$$

suppose the function $f(z)$ has p zeros at the point a_k ($k = 1, ..., p$) (roots are counted according to their multiplicity), and let

$$F(z) = f(z) \cdot \prod_{1}^{p} \frac{1-\bar{a}_k z}{a_k - z}.$$

We prove now that in $\tilde{D}_{r\varrho}$ there exists a point z^* such that

$$|F(z^*)| \leq \varepsilon \left[\frac{1}{C_r \mu_p^{r\varrho}(A)} \right]^p \qquad (2.44)$$

Let z be some point of $\tilde{D}_{r\varrho}$, and z_r be the radial projection z onto $\tilde{\Gamma}_r$. By virtue of Lemma 1 we have the inequality

$$\left| \prod_1^p \frac{1 - \bar{a}_k z}{a_k - z} \right| \leq \left| \prod_1^p \frac{1 - \bar{a}_{kr} z_r}{a_{kr} - z_r} \right| \qquad (2.45)$$

Let us consider now the set γ_r. Since the function $f(z)$ has p zeros in $\bar{D}_{r\varrho}$, it follows that the set γ_r consists of no more than p intervals. Then by virtue of

$$\gamma_r \supset A_{wr}$$

it follows that the measure of γ_r satisfies the inequality

$$\mu(\gamma_r) \geq \mu_p^{r\varrho}(A).$$

We denote by γ_r' the part of γ_r which belongs to the semi-circle $0 \leq \arg z \leq \pi$, and by γ_r'' the difference

$$\gamma_r'' = \gamma_r - \gamma_r'$$

It is evident that one of the two inequalities holds

$$\mu(\gamma_r') \geq \frac{1}{2} \mu(\gamma_r);$$

$$\mu(\gamma_r'') \geq \frac{1}{2} \mu(\gamma_r).$$

For the sake of definiteness assume

$$\mu(\gamma_r') \geq \frac{1}{2} \mu(\gamma_r).$$

By virtue of Lemma 2, in proving (2.44) one may consider γ_r to consist of one interval. From CHEBYSHEFF's results on polynomials which deviate least from zero it follows that on the interval γ_r' there is a point z_r^* such that

$$\prod_1^p |a_{kr} - z_r^*| \geq [C_r' \mu(\gamma_r')]^p \geq [C_r' \mu_p^{r\varrho}(A)]^p, \qquad (2.46)$$

where C_r is some constant which depends on r. From (2.46) and (2.45) we see that on the set γ we can find a point z^* such that

$$\left|\prod_1^p \frac{1-a_k z^*}{a_k - z^*}\right| \le \left[\frac{1}{C_r \mu_p^{r\varrho}(A)}\right]^p, \tag{2.47}$$

and the inequality to be proved follows from (2.44).

Now we consider the function

$$v(z) = \ln |F(z)|$$

The function $F(z)$ has no zeros in the annulus $\bar{D}_{r\varrho}$ and therefore the function $v(z)$ is a regular harmonic function in $\bar{D}_{r\varrho}$. Then by virtue of (2.44) and the conditions of the theorem

$$v(z^*) \le -p \ln \mu_p^{r\varrho}(A) + \ln \varepsilon$$
$$v(z) \le 0 \quad (z \in \bar{D}_{r\varrho}) \tag{2.48}$$

We denote by $\bar{\Gamma}_{r\varrho}$ the boundary of the annulus $\bar{D}_{r\varrho}$. It follows from (2.48) that

$$\int_{\tilde{\Gamma}_{r\varrho}} v(z)\, ds \le C'_{r\varrho}\left[\ln \varepsilon - p \ln \mu_p^{r\varrho}(A)\right] \tag{2.49}$$

where $C'_{r\varrho}$ is a known constant dependent on the radii r, ϱ and it follows from (2.49) that the function $v(z)$, in the annulus $D_{r\varrho}$, satisfies the inequality

$$v(z) \le C''_{r\varrho}\left[\ln \varepsilon - p \ln \mu_p^{r\varrho}(A)\right] \tag{2.50}$$

From the inequality (2.50) we obtain

$$|f(z)| \le \left\{\frac{\varepsilon}{[\mu_p^{r\varrho}(A)]^p}\right\}^{C''_{r\varrho}} \tag{2.51}$$

It is not difficult to obtain from the inequality (2.51) the following estimate for the modulus $|f(z)|$ in the whole disc D

$$|f(z)| \le \left\{\frac{\varepsilon}{[\mu_p^{r\varrho}(A)]^p}\right\}^{C''_{r\varrho}(1-|z|)} \tag{2.52}$$

Then, since the function $f(z)$ has p zeros in the annulus $\bar{D}_{r\varrho}$ it follows that

$$|f(z)| \le v^p, \tag{2.53}$$

where

$$v = \frac{|z| + \tilde{\varrho}}{1 + |z| \cdot \tilde{\varrho}}; \quad \tilde{\varrho} = \frac{2 + \varrho}{3}.$$

It is easy to see that

$$\ln v < C_\varrho \cdot (1 - |z|) \tag{2.54}$$

where C_ϱ is a constant dependent on ϱ.

Let p satisfy (2.34). Then by virtue of (2.52), (2.53), and (2.54) the inequality

$$|f(z)| \leq \exp\left\{-C_1(1-|z|)\frac{\ln \varepsilon}{\ln \mu_p^{r\varrho}(A)}\right\}$$

is valid, and this concludes the proof.

We cite some examples of obtaining $\mu_n^{r\varrho}(A)$ and the corresponding estimates.

1. Let the set A_{wr} have positive JORDAN measure $\mu(A_{wr})$. Then, it is evident that

$$\mu_n^{r\varrho}(A) \geq \mu(A_{wr});$$

$$\mu_n^{r\varrho}(A) \underset{n \to \infty}{\to} \mu(A_{wr}).$$

If, in particular, A_{wr} consists of m curves, then

$$\mu_n^{r\varrho}(A) > \mu(A_{wr}), \qquad n < m,$$

$$\mu_n^{r\varrho}(A) = \mu(A_{wr}), \qquad n \geq m.$$

If $m = 1$, then the estimate of the theorem, accurate up to the constant $C_{r\varrho}$, coincides with the estimates for an analytic function using harmonic measure.

2. Let the set A consist of the points a_k ($k = 1, \ldots, n$)

$$|a_k| = a.$$

$$\arg a_k = k\delta, \qquad \left(\delta = \frac{\pi}{n}\right).$$

One may easily see that in this case

$$\mu_p(A_r) = a(n-p)\delta, \qquad p \leq n,$$

$$\mu_p(A_r) = 0, \qquad p > n.$$

The relations (2.34) in this case are as follows

$$[(n-p)\delta \cdot a]^p \leq \varepsilon^{C'_a} \leq [(n-p+1)\delta \cdot a]^{p-1} \tag{2.55}$$

It is not difficult to show that the number p satisfying (2.55) satisfies the inequality

$$p \geq C'_a \ln \varepsilon' \frac{n\delta}{1 + \delta' \ln \varepsilon'}$$

from which we obtain

$$|f(z)| \leq \exp\left\{C_a \frac{n\delta' \ln \varepsilon'}{1 + \delta' \ln \varepsilon'}(1-|z|)\right\} \tag{2.56}$$

3. Let the set A consist of the points a_k $(k = 1, ..., \infty)$

$$|a_k| = a, \qquad \arg a_k = \frac{\pi}{2^k}.$$

One can see in this case that

$$\mu_p(A) = \frac{a}{2^p} \qquad\qquad (2.57)$$

and the relations (2.34) are as follows

$$\left(\frac{a}{2^p}\right)^p \leq \varepsilon^{C'_a} \leq \left(\frac{a}{2^{p-1}}\right)^{p-1}$$

from which

$$p^* \geq C''_a \sqrt{|\ln \varepsilon|}$$

and

$$|f(z)| \leq \exp\left\{-C_a \sqrt{.\ln|\,\varepsilon|} \cdot (1 - |z|)\right\}. \qquad\qquad (2.58)$$

§ 4. Analytic Continuation of a Function of Two Real Variables

Let $f(x, y)$ be a function of the two real variables, analytic and regular in the disc

$$x^2 + y^2 \leq 1, \qquad (D) \qquad\qquad (2.59)$$

and let A be some set inside D. Let the values of the function $f(x, y)$ be known on the set A and suppose it is required to determine $f(x, y)$ inside D. We prove a theorem characterizing the stability of the solution for a certain class of sets A.

First some definitions are introduced. Let us divide the x, y-plane into a system of parallel strips H_k of width h_k $(k = \pm 1, ..., \pm \infty)$. Let T denote this division while A_k denotes the part of the set A that belongs to H_k

$$A_k = A \cap H_k$$

By $\bar{A}_k$ we denote the projection of the set A_k onto the center line of the strip H_k. Let $\tau > 0$, C_j $(j = 1, ...)$ be some constants and let n_k be an integer satisfying the inequalities

$$\left[\frac{\mu_{n_k+1}(\bar{A}_k)}{C_1}\right]^{n_k+1} \leq \tau + C_2 h_k \leq \left[\frac{\mu_{n_k}(\bar{A}_k)}{C_1}\right]^{n_k} \qquad\qquad (2.60)$$

where $\mu_n(\bar{A}_k)$ is the n-measure of the set $\bar{A}_k$.

We denote by $H_{\lambda\tau}(A, T)$ a set of strips H_k for which the following inequality is true

$$\frac{\ln(\tau + C_2 h_k)}{C_3 \ln \mu_{n_k}(\bar{A}_k)} \geq \lambda, \tag{2.61}$$

and by $\mu_{m\lambda\tau}(A, T)$ the m-measure of the projection of the set $H_{\lambda\tau}(A, T)$ onto some straight line perpendicular to the lines of the division T. We call the supremum of $\mu_{m\lambda\tau}(A, T)$, for all possible divisions T, the (m, λ, τ)-measure of the set A and we denote it by

$$\mu_{m\lambda\tau}(A) = \sup_T \{\mu_{m\lambda\tau}(A, T)\}.$$

Theorem. Let A be a set of points of the x, y-plane lying in the disc

$$x^2 + y^2 \leq R^2 < 1 \qquad (D_R) \tag{2.62}$$

and let $f(z_1, z_2)$ be an analytic function of two complex variables z_1, z_2 regular in the hypercylinder

$$\begin{aligned} (\operatorname{Re} z_1)^2 + (\operatorname{Re} z_2)^2 &\leq 1, \\ |\operatorname{Im} z_1| \leq \varkappa, \quad |\operatorname{Im} z_2| &\leq \varkappa. \end{aligned} \qquad (\Omega)$$

If the function $f(z_1, z_2)$ satisfies the inequalities

$$\begin{aligned} |f(z_1, z_2)| &\leq 1, & (z_1, z_2) \in \Omega, \\ |f(x, y)| &\leq \varepsilon, & (x, y) \in A, \end{aligned} \tag{2.63}$$

then the inequality

$$|f(x, y)| \leq \exp\left\{\frac{-C_1 \lambda}{\ln \lambda_{m\lambda\varepsilon}(A)}\right\}, (x, y) \in D_R \tag{2.64}$$

is valid, where the numbers m, λ satisfy the relations

$$\left|\frac{\mu_{m+1, \lambda, \varepsilon}(A)}{C_2}\right|^{m+1} \leq e^{-\lambda} \leq \left|\frac{\mu_{m\lambda\varepsilon}(A)}{C_2}\right|^m \tag{2.65}$$

and C_j are constants depending on $R, \varkappa$.

We denote by T_α the division of the plane x, y for which

$$|\mu_{m\lambda\varepsilon}(A) - \mu_{m\lambda\varepsilon}(A, T)| < \alpha. \tag{2.66}$$

One can obviously consider the corresponding strips H_k to be parallel to the x-axis. Let the center line of the strip H_k be the straight line

$$y = y_k$$

It is not difficult to show that from (2.63) it follows that

$$|\text{grad } f(x, y)| \le C_3, \qquad (x, y) \in D_R \tag{2.67}$$

where C_3- depends on R and $\varkappa$. From (2.67) and (2.63) we find that at the points of the x, y-plane belonging to the set $\bar{A}_k$ the function $f(x, y)$ satisfies the inequality

$$|f(x, y)| \le \varepsilon + \frac{1}{2} C_3 h_k. \tag{2.68}$$

Let us consider the function

$$\varphi_k(z) = f(z, y_k), \qquad (x, y) \in H_k.$$

From the condition of the theorem and from (2.68) it follows that the function $\varphi_y(z)$ is an analytic function, regular in the domain

$$\begin{array}{c} |\text{Re } z| < \sqrt{1 - y_k^2}; \\ |\text{Im } z| < \varkappa; \end{array} \qquad (\Omega_k')$$

and satisfies the inequalities

$$\begin{array}{ll} |\varphi_k(z)| \le 1, & z \in \Omega_k'; \\ |\varphi_k(z)| \le \varepsilon + C_3 h_k, & (\text{Re } z, y_k) \in \bar{A}_k. \end{array} \tag{2.69}$$

Applying the results of the preceding section to the function $\varphi_y(z)$ we find that the following inequality is valid

$$|\varphi_k(z)| \le \exp \left\{ \frac{\ln(\varepsilon + C_3 h_k) \cdot C_4}{\ln \mu_{n_k}(\bar{A}_k)} \right\}, \qquad \begin{array}{l} |\text{Re } z| \le \dfrac{1+R}{2} \\[2mm] |\text{Im } z| < \dfrac{\varkappa}{2} \end{array}$$

where the constants C_3 and C_4 depend on R, $\varkappa$ hence

$$|f(x, y)| \le \exp \left\{ \frac{\ln(\varepsilon + C_3 h_k) \cdot C_4}{\ln \mu_{n_k}(\bar{A}_k)} \right\}, \qquad (x, y) \in H_k. \ \cap D_R \tag{2.70}$$

Now we consider the function

$$\psi_x(z) = f(x, z).$$

From the conditions of the theorem, the definition of the set $H_{\lambda\tau}(A, T)$ and (2.68) it follows that the function $\psi_x(z)$ is analytic, regular in the domain

$$\begin{array}{c} |\text{Re } z| < \sqrt{1 - x^2} \\ |\text{Im } z| < \varkappa \end{array} \qquad (\Omega_x'')$$

and satisfies the following inequalities

$$|\psi_x(z)| \leq 1, \qquad z \in \Omega_x'',$$

$$(2.71)$$

$$|\psi_x(y)| \leq e^{-\lambda}, \qquad (x, y) \in H_{\lambda\tau}(A, T_\alpha) \cap D_R.$$

Applying, as previously, the results of the preceding section we obtain, by virtue of (2.71)

$$(x, \operatorname{Re} z) \in D_R$$

$$|\psi_x(z)| \leq \exp\left\{\frac{\lambda C_5}{\ln \mu_{m\lambda\varepsilon}(A, T_\alpha)}\right\}$$

$$|\operatorname{Im} z| < \frac{\varkappa}{2}$$

from which it follows that

$$|f(x, y)| \leq \exp\left\{\frac{\lambda \cdot C_5}{\ln \mu_{m\lambda\varepsilon}(A, T_\alpha)}\right\}, \qquad (x, y) \in D_R \qquad (2.72)$$

Passing to the limit at $\alpha \to 0$ in the inequality (2.72) we obtain the inequality (2.64) to be proved.

Now we give two examples of the applicability of the theorem.

1) Let A be a closed set with positive plane measure $\mu(A)$. We show that in this case when

$$\lambda \leq \frac{|\ln \tau| \cdot C}{|\ln \mu(A)|}$$

the inequality

$$\mu_{m\lambda\varepsilon}(A) \geq \frac{1}{8}\mu(A) \qquad (2.73)$$

is valid, and consequently

$$|f(x, y)| \leq \exp\left\{-C\frac{\ln \varepsilon}{\ln \mu(A)}\right\} \qquad (2.74)$$

where C is some constant.

Let N be some number. We perform a division T of the plane into strips

$$\frac{k-1}{N} \leq y \leq \frac{k}{N} \qquad (H_k)$$

and denote by H_{kj} the square

$$\frac{k-1}{N} \leq y \leq \frac{k}{N}$$

$$(H_{kj})$$

$$\frac{j-1}{N} \leq x \leq \frac{j}{N}$$

We denote by p_k the number of squares H_{kj} of the strip H_k containing points of A, and by p the total number of squares containing points of A

$$p = \sum_{-\infty}^{\infty} p_k .$$

One can easily see that the inequalities

$$p \geq \mu(A) \cdot N^2$$

$$\mu_m(\bar{A}_k) \geq \frac{p_k - 2m}{N}$$

(2.75)

are valid. In fact, the first of the inequalities (2.75) follows from the definition of plane measure of the closed sets. In order to obtain the second of the inequalities (2.75) we note that by virtue of the definitions of p_k and $\bar{A}_k$, on the straight line

$$y = y_k = \frac{k - 0.5}{N}$$

there are p_k non-overlapping intervals of length $1/N$ containing points of $\bar{A}_k$. Thus the sum of the lengths of any intervals covering $\bar{A}_k$ cannot be less than the number

$$\frac{p_k - 2m}{N}$$

from which the second inequality follows.

Let us consider the strips H_k for which the inequality

$$p_k \geq \frac{1}{4} N \cdot \mu(A)$$

(2.76)

is valid.

Let (2.76) hold for

$$k = k_j \qquad (j = 1, ..., q)$$

It is not difficult to show that the number q satisfies the inequality

$$q \geq \frac{1}{4} N \cdot \mu(A) .$$

(2.77)

In fact, assuming the opposite we see that

$$p < 2Nq + (2N - q) \cdot \frac{1}{4} N \cdot \mu(A) < N^2 \cdot \mu(A)$$

which contradicts (2.75).

We denote by $\bar{H}$ the set of center lines of the strips H_{kj} ($j = 1, ..., q$), and by $\tilde{H}$ the projection of $\bar{H}$ into the straight line $x = 0$.

In analogy with the second of the inequalities (2.75) we find that for any n, the n-measure of $\tilde{H}$ satisfies the inequality

$$\mu_n(\tilde{H}) \geq \frac{q - 2n}{N}$$

from which, by virtue of (2.75) and the definition of $\mu_{n\lambda\tau}(A, T)$, we obtain

$$n < \frac{1}{16} N \cdot \mu(A)$$

$$\mu_{n\lambda\tau}(A, T) \geq \frac{1}{8} \mu(A) \qquad (2.78)$$

$$\lambda > \frac{\ln \tau \cdot C}{\ln \mu(A)}$$

Passing to the limit as $N \to \infty$ in (2.78) we obtain the inequality (2.64).

2) Let the set A consist of points p_{ij} with coordinates

$$x_{ij} = \frac{1}{2^j}, \qquad i = 1, ..., \infty$$

$$y_{ij} = \frac{1}{2^i} \qquad j = i, i+1, ..., \infty$$

We show that in this case for $\lambda < C_3 \sqrt{\ln \tau}$ the following inequality is valid

$$\mu_{m\lambda\tau}(A) \geq \frac{1}{2^m} - \frac{1}{2^{\sqrt{|\ln \tau|}}} \qquad (2.79)$$

and consequently,

$$|f(x, y)| \leq \exp\{-C|\ln \varepsilon|^{1/4}\} \qquad (2.80)$$

Let us perform the division T of the plane x, y into the strips H_k

$$\frac{k-1}{2^N} \leq y \leq \frac{k}{2^N}$$

where N is some number.

It is easy to see that for $k = 2^p$ ($p = 1, ..., N - 1$) the n-measure of the set $\bar{A}_k$ is equal to

$$\mu_n(\bar{A}_k) = \frac{1}{2^{n+p}}. \qquad (2.81)$$

Therefore the relation (2.60) in this case are as follows

$$\left(\frac{C_1}{2^{n_k+p+1}}\right)^{n_k+1} \le \tau_1 \le \left(\frac{C_1}{2^{n_k+p}}\right)^{n_k}, \tag{2.82}$$

where

$$\tau_1 = \tau + \frac{C_2}{2^N}.$$

From (2.82) it follows that for

$$p \le \sqrt{|\ln \tau_1|}, \qquad \lambda \le C_3 \sqrt{|\ln \tau_1|}$$

the straight line

$$y = y_k = \frac{k - 0.5}{2^N}, \qquad (k = 2^p)$$

belongs to the set $H_{\lambda\tau}(A, T)$.

By virtue of the above statement, one can show that the measure of the projection of $H_{\lambda\tau}(A, T)$ onto the straight line $x = 0$ satisfies the inequality

$$\mu_m[H_{\lambda\tau}(A, T)] \ge \frac{1}{2^m} - \frac{1}{2^{\sqrt{|\ln \tau_1|}}}$$

from which follows

$$\mu_{m\lambda\tau}(A, T) \ge \frac{1}{2^m} - \frac{1}{2^{\sqrt{|\ln \tau_1|}}} \tag{2.83}$$

Passing to the limit as $N \to \infty$ in (2.83) we obtain the inequality (2.79).

§ 5. Analytic Continuation of Harmonic Functions from a Circle

Let $u(z)$ be a harmonic function regular and bounded in the disc

$$|z| \le 1 \qquad (D)$$

and let its value be known on the circle

$$|z| = r < 1 \qquad (\Gamma_r),$$

i.e.,

$$u(re^{i\varphi}) = f(\varphi)$$

where $f(\varphi)$ is a known function. It is required to determine the function $u(z)$ inside the disc $|z| < r$ from the values of the function on Γ_r. Let us prove a theorem characterizing the stability of the solution.

Theorem 1. Let the considered function $u(z)$ satisfy the inequalities

$$\frac{1}{\pi} \int_{-\pi}^{\pi} u^2 (re^{i\varphi}) \, d\varphi \le \varepsilon,$$

$$\frac{1}{\pi} \int_{-\pi}^{\pi} u^2 (e^{i\varphi}) \, d\varphi \le 1.$$

(2.84)

Then the following inequality is valid

$$\frac{1}{\pi} \int_{-\pi}^{\pi} u^2 (\varrho e^{i\varphi}) \, d\varphi \le \exp \left\{ \frac{\ln \varepsilon \cdot \ln \varrho}{\ln r} \right\}$$

(2.85)

It is well known that

$$u(\varrho e^{i\varphi}) = \sum_{1}^{\infty} \varrho^k (a_k \cos k\varphi + b_k \sin k\varphi) + \frac{a_0}{2}.$$

Let us write

$$\sqrt{a_k^2 + b_k^2} = C_k, \qquad \sqrt{\frac{a_0}{2}} = C_0.$$

From (2.84) we have

$$\sum_{0}^{\infty} r^{2k} C_k^2 \le \varepsilon,$$

$$\sum_{0}^{\infty} C_k^2 \le 1.$$

(2.86)

Let us estimate the function

$$\omega(\varrho) = \sum_{0}^{\infty} \varrho^{2k} C_k^2 = \frac{1}{\pi} \int_{-\pi}^{\pi} u^2 (\varrho e^{i\varphi}) \, d\varphi$$

using (2.86).

By virtue of (2.86) the sum to be estimated is bounded. Let the sum reach its conditional maximum over the variables C_k for $C_k = \bar{C}_k$ ($k = 0, ..., \infty$). It is evident that

$$\bar{C}_k = 0, \qquad k \ne p, q,$$

where $p, q, p < q$, are some numbers, since otherwise it is quite clear that for $\bar{C}_k$ one may choose variations δ_k such that

$$\sum_{0}^{\infty} (\bar{C}_k + \delta_k)^2 \varrho^{2k} \ge \sum_{0}^{\infty} \bar{C}_k^2 \varrho^{2k};$$

$$\sum_{0}^{\infty} (\bar{C}_k + \delta_k)^2 < \sum_{0}^{\infty} \bar{C}_k^2;$$

$$\sum_{0}^{\infty} (\bar{C}_k + \delta_k)^2 r^{2k} < \sum_{0}^{\infty} \bar{C}_k^2 r^{2k}.$$

Therefore $\bar{C}_p$, $\bar{C}_q$ satisfy one of the three relations

$$\left.\begin{array}{l} \bar{C}_p^2 + \bar{C}_q^2 = 1, \\ \bar{C}_p^2 r^{2p} + \bar{C}_q^2 r^{2q} = \varepsilon; \end{array}\right\} (I)$$

$$\bar{C}_p = 0; \qquad (II)$$
$$\bar{C}_q = 0. \qquad (III)$$

(2.87)

Let the first of the relations (2.87) hold. Then

$$\bar{C}_p^2 = \frac{\varepsilon - r^{2q}}{r^{2p} - r^{2q}};$$

$$\bar{C}_q^2 = \frac{r^{2p} - \varepsilon}{r^{2p} - r^{2q}}.$$

(2.88)

We note that from (2.88) follows

$$r^{2q} \le \varepsilon \le r^{2p}. \qquad (2.89)$$

By virtue of (2.87), (2.88) we obtain

$$\sum_0^\infty \bar{C}_k^2 \varrho^{2k} = \varepsilon \frac{\varrho^{2p} - \varrho^{2q}}{r^{2p} - r^{2q}} + \frac{r^{2p}\varrho^{2q} - r^{2q}\varrho^{2p}}{r^{2p} - r^{2q}}. \qquad (2.90)$$

It is easy to show that for $p < q$ the functions

$$F_1(p,q) = \frac{\varrho^{2p} - p^{2q}}{r^{2p} - r^{2q}};$$

$$F_2(p,q) = \frac{r^{2p}\varrho^{2q} - r^{2q}\varrho^{2p}}{r^{2p} - r^{2q}}$$

increase with increasing p as well as with decreasing q from which, by virtue of (2.89) and (2.90), it follows that

$$\sum_0^\infty \bar{C}_k^2 \varrho^{2k} \le \lim_{p,\,q \to \frac{\ln \varepsilon}{\ln r} \cdot \frac{1}{2}} \{\varepsilon F_1(p,q) + F_2(p,q)\} = \exp\left\{\frac{\ln \varepsilon \cdot \ln \varrho}{\ln r}\right\}. \qquad (2.91)$$

Now let the second of the relations (2.87) hold. Then

$$\sum_0^\infty \bar{C}_k^2 \varrho^{2k} = C_q^2 \varrho^{2q}$$

and by virtue of (2.86)

$$\bar{C}_q^2 \le 1,$$
$$\bar{C}_q^2 r^{2q} \le \varepsilon,$$

from which we obtain

$$\sum_0^\infty \bar{C}_k^2 \varrho^{2k} \leq \exp\left\{\frac{\ln \varepsilon \cdot \ln \varrho}{\ln r}\right\}. \tag{2.92}$$

In case the third of the relations (2.87) holds, inequality (2.92) can be obtained in a similar manner.

The statement of the theorem follows from (2.91), (2.92) and from the definition of the numbers $\bar{C}_k$.

With respect to the function $u(z)$ considered in theorem 1, let it be known that at the boundary of the domain the n-th derivative is bounded in L_2, i.e.,

$$\frac{1}{\pi} \int_{-\pi}^\pi \left[\frac{\partial^n}{\partial \varphi^n} u(e^{i\varphi})\right]^2 d\varphi \leq M$$

and suppose it is required to determine the function $u(z)$ on the circle $\varrho = 1$ from the values of the function $u(z)$ on the circle $\varrho = r$. We shall prove a theorem characterizing the stability of the solution.

Theorem 2. If the function $u(z)$ satisfies the relation

$$\frac{1}{\pi} \int_{-\pi}^\pi u^2(re^{i\varphi}) d\varphi \leq \varepsilon;$$

$$\frac{1}{\pi} \int_{-\pi}^\pi \left[\frac{\partial^n}{\partial \varphi^n} u(e^{i\varphi})\right]^2 d\varphi \leq 1; \tag{2.93}$$

$$\int_{-\pi}^\pi u(e^{i\varphi}) d\varphi = 0,$$

then the following inequality is valid

$$\frac{1}{\pi} \int_{-\pi}^\pi u^2(e^{i\varphi}) d\varphi \leq \frac{1}{\lambda^{2n}}, \tag{2.94}$$

where λ is a root of the equation

$$2\lambda \ln r - 2n \ln \lambda = \ln \varepsilon. \tag{2.95}$$

We note that for a sufficiently small ε the root of equation (2.95) satisfies the inequality

$$\lambda > \frac{1}{4} \frac{\ln \varepsilon}{\ln r}. \tag{2.96}$$

The proof of theorem 2 is similar to that of theorem 1. From (2.93) it follows that

$$\sum_0^\infty r^{2k} C_k^2 \le \varepsilon;$$

$$\sum_0^\infty k^{2n} C_n^2 \le 1,$$

(2.97)

where C_k, as in the previous theorem, are expressed by the expansion coefficients of the function $u\,(\varrho e^{i\varphi})$ in a FOURIER series in the polar angle φ.

In analogy with (2.87) we find that the numbers $\bar{C}_k$ for which the sum

$$\sum_0^\infty C_k^2 = \frac{1}{\pi} \int_{-\pi}^{\pi} u^2 (e^{i\varphi})\,d\varphi$$

attains its conditional maximum, satisfy the relations

$$\bar{C}_k = 0, \qquad k \ne p, q,$$

while $\bar{C}_p, \bar{C}_q$ satisfy one of the three relations

$$\left.\begin{array}{l} r^{2p}\bar{C}_p^2 + r^{2q}\bar{C}_q^2 = \varepsilon \\ p^{2n}\bar{C}_p^2 + q^{2n}\bar{C}_q^2 = 1 \end{array}\right\} \text{(I)} \qquad \begin{array}{l} \bar{C}_p = 0 \quad \text{(II)} \\ \bar{C}_q = 0 \quad \text{(III)} \end{array}$$

from which the theorem follows in an analogous manner.

§ 6. Analytic Continuation of Harmonic Functions with Cylindrical Symmetry

Let x, y be vectors with components $(x_1, \ldots, x_n)$, $(y_1, \ldots, y_{m+1})$ and let $u\,(x, y)$ be a harmonic function regular and continuous in the domain

$$|x| \le 1, \qquad |y| \le 1 \qquad (D)$$

(2.98)

Let also the function $u\,(x, y)$ as a function y depend on the variable $r = |y|$ alone. It is required to determine the values of the function $u\,(x, y)$ inside D from its values on the set P

$$x, y \in P, \quad \text{if} \quad y = 0; \qquad |x| \le \varrho < 1.$$

We shall prove a theorem characterizing the stability of the solution.

Theorem. If the function $u\,(x, y)$ to be considered satisfies the inequalities

$$|u\,(x, y)| \le 1, \qquad (x, y) \in D;$$

$$|u\,(x, y)| \le \varepsilon, \qquad (x, y) \in P$$

(2.99)

then the following inequality also holds

$$|u(x, y)| \leq C\varepsilon^{\gamma}, \qquad (x, y) \in D_1,$$

where D_1 is any domain lying strictly inside D and C, γ are some constants dependent on D_1 and ϱ.

By virtue of the conditions of the theorem, the function $u(x, y)$ may be represented as a function $u(x, r)$ where $r = |y|$. The function $u(x, r)$ will satisfy equation

$$\frac{\partial^2 u}{\partial r^2} + \frac{m}{r}\frac{\partial u}{\partial r} + \Delta_x u = 0. \qquad (2.100)$$

From the known properties of harmonic functions it follows that $u(x, r)$ is analytic with respect to the variables x, r everywhere inside D.

We will estimate the modulus of the function $\Delta_x^p u(x, 0)$. We consider the function $f(z)$ where z is a vector with complex components $z_k = x_k + i\xi_k$ ($k = 1, \ldots, n$) analytic with respect to the variables z_k and coinciding on a real hyperplane with the function $u(x, 0)$

$$f(x) = u(x, 0).$$

It is evident that the function $f(z)$ is regular and bounded in the domain R of the n-dimensional complex space

$$z \in R, \quad \text{if} \quad |\operatorname{Re} z| \leq \varrho, \quad |\operatorname{Im} z| \leq h, \quad \varrho < \varrho_1 < 1,$$

where h is a constant dependent on ϱ_1. Define R_1 to be the domain

$$z \in R_1 \quad \text{if} \quad |\operatorname{Re} z| \leq \varrho, \quad |\operatorname{Im} z| \leq h_1 < h$$

From (2.99) as well as from the results of § 1 it follows that the function $f(z)$ in the domain R_1 satisfies the inequality

$$|f(z)| \leq C_1 \varepsilon^{\gamma_1}; \qquad (2.101)$$

where C_1, γ_1 are some constants dependent on ϱ, h, ϱ_1, h_1 whereas from (2.101) and from known properties of analytic functions it follows that

$$|f^{(k)}(z)| \leq k! \alpha^k \varepsilon^{\gamma_1} \qquad (2.102)$$

where α is a constant.

From (2.102) we obtain

$$\Delta_x^p u(x, 0) \leq (2p)! \alpha_1^k \varepsilon^{\gamma_1} \qquad (2.103)$$

We now expand the function $u(x, r)$ in a MACLAURIN series in the variable r

$$u(x, r) = \sum_0^\infty \frac{1}{(2p)!} \frac{\partial^{2p}}{\partial r^{2p}} u(x, 0) \cdot r^{2p} \qquad (2.104)$$

(the coefficients of the odd powers of r are zero by virtue of the symmetry).

Substituting (2.104) in (2.100) we find that the functions $\partial^{2p}/\partial r^{2p}\, u\,(x, 0)$ satisfy the following relations

$$\frac{\partial^{2p}}{\partial r^{2p}} u\,(x, 0) = -\frac{2\,p-1}{2\,p-1+m}\, \varDelta\, \frac{\partial^{2\,(p-1)}}{\partial r^{2\,(p-1)}} u\,(x, 0) \quad (p=1, ..., \infty)$$

from which

$$\frac{\partial^{2p}}{\partial r^{2p}} u\,(x, 0) = (-1)^p \frac{(2\,p-1)!!\,(m-1)!!}{(2\,p-1+m)!!}\, \varDelta^p u\,(x, 0) \qquad (2.105)$$

The validity of the statement of the theorem for any neighbourhood P follows from (2.104), (2.105) and (2.103). The validity of the assertion of the theorem for the arbitrary domain D_1 follows from the validity of this assertion for the neighbourhood P and from general estimates for analytic continuations.

Chapter III

Inverse Problems for Differential Equations

We consider two types of inverse problems for linear differential equations.

The first kind is the so-called inverse problem of potential theory where the equation is non-homogeneous, the coefficients of the differential operator are given and it is required to determine the right-hand side. In this case we shall restrict our consideration to the LAPLACE operator, although a number of results are carried over to more general elliptic equations.

The second kind is that in which the equation is homogeneous and it is required to determine some of its coefficients. A well-known problem of this kind is the inverse problem for the STURM-LIOUVILLE equation.

§ 1. The Inverse Problem for a Newtonian Potential

By an inverse problem for a NEWTONIAN potential we mean the following. On a part of the surface of the unit sphere D, there is known a harmonic function $u(x)$ which is an external NEWTONIAN potential of a body with unit density which is star-shaped with respect to some point η and lies inside the sphere

$$|x| \leq R < 1 \quad (D_R)$$

It is required to determine the dimensions and shape of the body from the function $u(x)$. This problem has an important application in geophysics and a number of papers are devoted to its investigation.

The uniqueness of the solution of the problem was proved by NOVIKOV. The notion of correctness in a compact set was introduced by TYKHONOV in an application to a problem. In [16] estimates were obtained which characterize as properly posed the problem for a class of domains bounded by continuously differentiable surfaces with bounded derivatives.

Let the polar coordinates of the vector $y - \eta$ be denoted by ϱ, φ, θ and let the surface of the body inducing the potential $u(x)$ be defined by the equation

$$\varrho = f(\varphi, \theta).$$

It is well-known that the function $u(x)$ is equal to

$$u(x) = \int_{-\pi/2}^{\pi/2} \int_{-\pi}^{\pi} \int_{0}^{f(\varphi,\theta)} \frac{\varrho^2 \sin \theta}{r(x, y)} d\varrho \, d\varphi \, d\theta \tag{3.1}$$

where $r(x, y)$ is the distance between the points x, y.

Equality (3.1) may be regarded as an integral equation for the unknown function $f(\varphi, \theta)$ which defines the body to be found. If the integral operator in the right-hand side of (3.1) is denoted by A then (3.1) will take the form

$$u(x) = Af(\varphi, \theta). \tag{3.1'}$$

The operator A in the right-hand side of (3.1') clearly transforms a function of the polar angles φ, θ to a function defined on some part of the unit sphere. Thus, the solution of the inverse problem of potential theory is equivalent to the solution of the nonlinear integral equation (3.1) or (3.1').

Now we proceed to the presentation of a stability theorem for the inverse problem of potential theory. For simplicity we restrict ourselves to the case in which the part of the surface D on which the potential $u(x)$ is known, is defined by the inequalities.

$$x_1 \geq H, \ |H| < 1 \quad (S_H)$$

Theorem 1. Let the functions $f_1(\varphi, \theta)$, $f_2(\varphi, \theta)$ satisfy the following conditions:

1) The functions $f_j(\varphi, \theta)$ $(j = 1, 2)$ are continuously differentiable for all values of the polar angles φ, θ and their derivatives satisfy the inequalities

$$\left| \frac{\partial}{\partial \theta} f_j(\varphi, \theta) \right| \leq M; \ \left| \frac{\partial}{\partial \varphi} f_j(\varphi, \theta) \right| \leq M(\pi^2 - \theta^2). \tag{3.2}$$

2) The following inequality holds

$$|u_1(x) - u_2(x)| \leq \varepsilon, \ x \in S_H \tag{3.3}$$

where

$$u_j(x) = Af_j(\varphi, \theta).$$

Then the difference $f_1(\varphi, \theta) - f_2(\varphi, \theta)$ satisfies the inequality

$$|f_1(\varphi, \theta) - f_2(\varphi, \theta)| \leq \frac{C_1}{n} \tag{3.4}$$

where C_1 is a constant depending on M, and n is a number defined by the relations

$$\left(\frac{1}{n+1} \right)^{2n+3} \leq C_2 \left| \ln \varepsilon \right|^{1/C_3} \leq \left(\frac{1}{n} \right)^{2n+1} \tag{3.5}$$

while C_j are constants depending on R, M.

We first prove a lemma.

Lemma 1. Let $\omega(y)$ be a summable function which does not exeed one in modulus and let the function $f(\varphi, \theta)$ satisfy the conditions of the problem as well as the inequalities (3.2). Then if the function

$$u(x)=\int\limits_{-\pi/2}^{\pi/2}\int\limits_{-\pi}^{\pi}\int\limits_{0}^{f(\varphi,\theta)}\frac{\varrho^2\sin\theta\cdot\omega(y)}{r(x,y)}\,d\varrho\,d\varphi\,d\theta$$

satisfies the inequality

$$|u(x)|\leq\varepsilon \quad (x\in S_H) \tag{3.6}$$

then the following inequality also holds

$$|u(x)|+|\operatorname{grad}u(x)|\leq C_4\,|\ln\varepsilon|^{-\frac{1}{c_5}} \tag{3.7}$$

where the constant C_j depends on R, M and x is any point lying outside or on the surface

$$\varrho=f(\varphi,\theta).$$

It may be shown from (3.6) that the function $u(x)$ satisfies the inequality

$$|u(x)|\leq\varepsilon^{C_6}=\varepsilon_1 \quad (|x|\geq 1) \tag{3.8}$$

where C_6 is a constant depending on the radii R and H.

Let the polar coordinates of the vector $x-\eta$ be denoted by $\theta',\varphi',\varrho'$ and the distance between the points x and y by

$$r(\varrho',\varphi',\theta',\varrho,\varphi,\theta)=r(x,y).$$

Let us consider the function

$$\omega(z,\varphi',\theta')=\int\limits_{-\pi/2}^{\pi/2}\int\limits_{-\pi}^{\pi}\int\limits_{0}^{f(\varphi,\theta)}\frac{\varrho^2\sin\theta\cdot\omega(y)}{r(z,\varphi',\theta',\varrho,\varphi,\theta)}\,d\varrho\,d\varphi\,d\theta.$$

It can be easily seen that $\omega(z,\varphi',\theta')$ is an analytic function of z which is regular in the domain

$$\operatorname{Re}z-C_7\,|\operatorname{Im}z|\leq f(\varphi',\theta') \quad (\Omega_{\varphi'\theta'})$$

where C_7 is a constant depending on the constant M in (3.2), and satisfies the following inequalities

$$|\omega(z,\varphi',\theta')|+|\operatorname{grad}\omega(z,\varphi',\theta')|\leq C_8,\ z\in\Omega_{\varphi',\theta'}\,; \tag{3.9}$$

$$\left|\frac{d}{dz}\operatorname{grad}\omega(z,\varphi',\theta')\right|\leq C_9\,|\ln[z-f(\varphi',\theta')]|, \tag{3.10}$$

$$\operatorname{Re}z\geq f(\varphi',\theta');\quad \operatorname{Im}z=0;$$

$$|\omega(z,\varphi',\theta')|\leq\varepsilon_1,$$

$$\operatorname{Re}z\geq 1,\quad \operatorname{Im}z=0, \tag{3.11}$$

where C_j are constants depending on M.

From the inequalities (3.9), (3.11), some properties of analytic functions and the results of Chapter II, it follows that the function $\omega\,(z,\varphi',\theta')$ satisfies the inequality

$$|\omega(z,\varphi',\theta')+|\operatorname{grad}\omega(z,\varphi',\theta')|\le\varepsilon^{C_{10}}$$

$$\operatorname{Re}z\ge1,\quad\operatorname{Im}z\le\frac{1}{2}C_7.\tag{3.12}$$

By virtue of (3.9), (3.10), and (3.12) the functions $\omega\,(z,\varphi',\theta')$ and grad ω satisfy the conditions of Carleman's theorem ([3] or [29]) and therefore the following inequality holds

$$|\omega(z,\varphi',\theta')|+|\operatorname{grad}\omega(z,\varphi',\theta')|\le C_{11}\exp\left\{\ln\varepsilon\,|z-f(\varphi',\theta')|^{C_{12}}\right\}.\tag{3.13}$$

Using inequalities (3.10) and (3.13) we obtain

$$
\begin{aligned}
|u(x)|+|\operatorname{grad}u(x)|&=|\omega(\varrho',\varphi',\theta')|+\\
&+|\operatorname{grad}\omega(\varrho',\varphi',\theta')|\le C_{11}\exp\left\{\ln\varepsilon\,[\bar{\varrho}-f(\varphi',\theta')]^{C_{12}}\right\}+\\
&+C_6[\bar{\varrho}-f(\varphi',\theta')]\cdot\{1+|\ln[\bar{\varrho}-f(\varphi',\theta')]|\};
\end{aligned}\tag{3.14}
$$

$$f(\varphi',\theta')\le\varrho'<\bar{\varrho}\le1+\psi.$$

Setting

$$\bar{\varrho}-f(\varphi',\theta')=|\ln\varepsilon|^{-\frac{1}{2C_{12}}}$$

and using the obvious inequality

$$|\ln[\bar{\varrho}-f(\varphi',\theta')]|\le C_{13}[\bar{\varrho}-f(\varphi',\theta')]^{\frac{1}{4}C_{12}}$$

we get the inequality (3.7).

Lemma 2. Let $\omega\,(x)$ be a harmonic function regular inside the surface

$$|x-\eta|=f(\varphi',\theta')+h,\qquad h>0\qquad(\Sigma_h)$$

where $f(\varphi,\theta)$ satisfies the conditions (3.2) and does not exceed one in modulus in this domain. Then on the surface

$$|x-\eta|=f(\varphi',\theta')\qquad\qquad(\Sigma_0)$$

the inequality

$$\sum_{j,k}\left|\frac{\partial^2\omega}{\partial x_j\partial x_k}\right|+\sum_j\left|\frac{\partial\omega}{\partial x_j}\right|\le\frac{C_1}{h^2}\tag{3.15}$$

holds, where the constant C_1 depends on M.

It is evident that by virtue of (3.2) for any point $x_0\in\Sigma_0$ there exists a sphere of radius C_2h with center at the point x_0 lying completely inside the surface Σ_h

(C_2 depends on M). By the maximum principle the values of $\omega(x)$ at the boundary of this sphere do not exceed one in modulus. The assertion of the Lemma follows from the above and from known estimates for derivatives of harmonic functions at the center of a sphere of given radius in terms of its values at the boundary of the sphere.

Lemma 3. Let $\omega(x)$ be a harmonic function regular inside the surface Σ_h of Lemma 2, satisfying the following conditions.

1) For some vector $x_0 \in \Sigma_h$ and for an arbitrary point x lying on the surface Σ_h and inside the sphere

$$|x - x_0| \le \lambda,$$

the function $\omega(x)$ is

$$\omega(x) = 1.$$

2) For an arbitrary $x \in \Sigma_h$

$$|\omega(x)| \le 1.$$

Then the following inequality holds

$$1 - \omega(x_{0h}) \le C_1 \left(\frac{h}{\lambda}\right)^{C_2} \tag{3.16}$$

where C_j are some constants depending on the constant M; $x_{0h} - \eta$ is the vector with the polar coordinates

$$\varrho_0 - h, \varphi_0, \theta_0$$

where $\varrho_0, \varphi_0, \theta_0$ are the polar coordinates of the vector x_0.

From the conditions imposed on the function $f(\varphi, \theta)$ and also from the maximum principle it follows that the assertion of the Lemma is equivalent to the following assertion.

Let $W'(x)$ be a harmonic function regular in the sphere with a cut-out cone

$$\varrho \le \lambda, \quad \frac{\pi}{2} - \theta \ge \alpha,$$

(ϱ, φ, θ are polar coordinates of $x - \eta$ and α is a constant depending on M). Suppose $W'(x)$ takes on the following boundary values

$$W'(x) = -1$$

on the surface of the sphere $\varrho = \lambda$;

$$W'(x) = 1$$

on the surface of the cone $\pi/2 - \theta = \alpha$. Then at the point x_h, where the polar coordinates of the vector $x_h - \eta$ are equal to

$$\varrho_h = h, \quad \varphi_h = 0, \quad \theta_h = -\frac{\pi}{2},$$

the function $W'(x)$ satisfies the inequality

$$1 - W'(x_h) \le C_1 \left(\frac{h}{\lambda}\right)^{c_2},$$

where C_j are constants depending on α.

The latter assertion follows from known estimates for harmonic functions.

Lemma 4. Let $\omega(t)$ be a nonincreasing function of the parameter t on the interval $[0, C]$

$$\omega(0) = 1, \quad \omega(C) = 0. \qquad (3.17)$$

Then for an arbitrary $t, 0 < t < 1, n \ge 1$ there exists a number $k, 1 \le k < n$ such that

$$\omega(t^{k+1}) - \omega(t^k) \le \frac{1}{n}. \qquad (3.18)$$

We consider the sum

$$\sum_1^{n-1} [\omega(t^{k+1}) - \omega(t^k)] = \omega(t^n) - \omega(t).$$

From the conditions of the lemma all the terms of the sum are non-negative and the whole sum does not exceed unity, from which it follows that some member of the sum does not exceed $1/n$, i.e., for some k (3.18) holds.

We now turn to the proof of the theorem. We introduce the following notations.

Let

$$f^l(\varphi, \theta) = \max_{j=1,2} [f_j(\varphi, \theta)],$$

$$f^i(\varphi, \theta) = \min_{j=1,2} [f_j(\varphi, \theta)].$$

We denote by Σ^l the surface defined by the equation

$$\varrho = f^l(\varphi, \theta)$$

and by Σ^i the surface defined by the equation

$$\varrho = f^i(\varphi, \theta);$$

and by $\Sigma^l_{(h)}$ the surface defined by the equation

$$\varrho = f^l(\varphi, \theta) + h.$$

We denote by $\Sigma^{l+}_{(h)}$ the part of the surface $\Sigma^{l}_{(h)}$ whose points satisfy the inequality

$$f_1(\varphi,\theta) \geq f_2(\varphi,\theta);$$

and by $\Sigma^{l-}_{(h)}$ we denote the part of $\Sigma^{l}_{(h)}$ whose points satisfy

$$f_1(\varphi,\theta) \leq f_2(\varphi,\theta).$$

The corresponding parts of the surfaces Σ^l, Σ^i will be denoted by

$$\Sigma^{l+}, \ \Sigma^{l-}, \ \Sigma^{i+}, \ \Sigma^{i-}.$$

We denote by $\Sigma^{l+}_{(h)\,\lambda}$ the part of $\Sigma^{l}_{(h)}$ whose distance from $\Sigma^{l}_{(h)}$ is not less than λ

$$r(x,\Sigma^{l-}_{(h)}) \geq \lambda \qquad x \in \Sigma^{l+}_{(h)\,\lambda}.$$

We denote by $\Sigma^{l}_{(h)\,\lambda}$ the part of $\Sigma^{l-}_{(h)}$ which belongs neither to $\Sigma^{l+}_{(h)\,\lambda}$, nor to $\Sigma^{l-}_{(h)\,\lambda}$, and finally, by $\Sigma^{l}_{(h)\,\lambda_1\,\lambda_2}$ we denote the part of $\Sigma^{l}_{(h)}$ which does not belong to the surfaces

$$\Sigma^{l+}_{(h)\,\lambda_2}, \ \Sigma^{l-}_{(h)\,\lambda_2}, \ \Sigma^{l}_{(h)\,\lambda_1} \quad (\lambda_2 > \lambda_1).$$

The parts of Σ^i, Σ^l to which correspond rays from the point η intersecting the indicated surfaces, will be denoted by

$$\Sigma^{i+}_{\lambda}, \ \Sigma^{i-}_{\lambda}, \ \Sigma^{i}_{\lambda}, \ \Sigma^{i}_{\lambda,\,\lambda_2}$$
$$\Sigma^{l+}_{\lambda}, \ \Sigma^{l-}_{\lambda}, \ \Sigma^{l}_{\lambda}, \ \Sigma^{l}_{\lambda_1\lambda_2}$$

respectively.

We denote by Ω the domain enclosed between the surfaces Σ^l, Σ^i while by Ω^+_{λ}, Ω^-_{λ} we denote the parts of Ω enclosed respectively between the surfaces Σ^{l+}_{λ}, Σ^{i+}_{λ} and Σ^{l-}_{λ}, Σ^{i-}_{λ}, and the radial straight lines bounding them.

Finally we denote by $\mu(\Omega)$ the volume of the domain Ω, by $\sigma(\Sigma)$ the area of the surface Σ, by $\alpha(x)$ the function defined by the relations

$$\alpha(x)=1 \qquad x \in \Omega^+_0;$$
$$\alpha(x)=-1 \qquad x \in \Omega^-_0;$$
$$\alpha(x)=0 \qquad \text{outside } \Omega;$$

and by $u(x)$ we denote the difference

$$u(x)=u_1(x)-u_2(x).$$

Clearly

$$u(x)=\int \frac{\alpha(\xi)}{r(x,\xi)} d\xi.$$

Let $v(x)$ be a harmonic function regular inside the surface Σ^l and continuously differentiable on Σ^l. Then from potential theory

$$\int_{\Omega} \alpha(x) \cdot v(x)\, dx = \frac{1}{4\pi} \int_{\Sigma^l} \left(\frac{\partial}{\partial n} u \cdot v - \frac{\partial}{\partial n} v \cdot u \right) d\sigma \qquad (3.19)$$

where $\partial/\partial n$ is the derivative along the normal to Σ^l.

We set

$$v(x) = \operatorname{div}\left[W(x)(x-\eta) \right]$$

where $W(x)$ is a harmonic function regular inside $\Sigma^l_{(h)}$ and assuming on $\Sigma^l_{(h)}$ the following boundary values

$$W(x) = 1 \qquad x \in \Sigma^{l+}_{(h)\frac{\lambda}{2}};$$

$$W(x) = -1 \qquad x \in \Sigma^{l-}_{(h)\frac{\lambda}{2}};$$

$$W(x) = 0 \qquad x \in \Sigma^{l}_{(h)\frac{\lambda}{2}}$$

We estimate first the right-hand side of (3.19). By virtue of Lemma 1 and the conditions of the theorem, the following inequality holds on the surface Σ^l

$$|u(x)| + |\operatorname{grad} u(x)| \le \varepsilon_1 = C_1 \, |\ln \varepsilon|^{-\frac{1}{C_2}}, \qquad (3.20)$$

and by virtue of Lemma 2

$$|v(x)| + |\operatorname{grad} v(x)| \le \frac{C_3}{h^2}. \qquad (3.21)$$

From (3.20), (3.21) and (3.2) we obtain

$$\left| \int_{\Sigma^l} \left[\frac{\partial}{\partial n} u \cdot v - \frac{\partial}{\partial n} v \cdot u \right] d\sigma \right| \le \frac{C_3}{h^2} \cdot \varepsilon_1 \cdot \sigma(\Sigma^l) \le \frac{C_4}{h^2} \cdot \varepsilon_1 \qquad (3.22)$$

where C_4 is a constant dependent on M.

We transform the left-hand side of (3.19) as follows

$$\int_{\Omega} \alpha(x) v(x)\, dx = \int_{\Omega} \alpha(x) \operatorname{div}\left[W(x)(x-\eta) \right] dx$$

$$= \int_{\Sigma^{i+}_{\lambda}} \alpha \cdot [W \cdot (x-\eta)]_n \, d\sigma - \int_{\Sigma^{i+}_{\lambda}} \alpha \cdot [W \cdot (x-\eta)]_n \, d\sigma$$

$$+ \int_{\Sigma^{i-}_{\lambda}} \alpha [W \cdot (x-\eta)]_n \, d\sigma - \int_{\Sigma^{i-}_{\lambda}} \alpha \cdot [W \cdot (x-\eta)]_n \, d\sigma$$

$$+ \int_{\Sigma^{l}_{h\lambda}} \alpha \cdot [W \cdot (x-\eta)]_n \, d\sigma - \int_{\Sigma^{i}_{h\lambda}} \alpha \cdot [W \cdot (x-\eta)]_n \, d\sigma$$

$$+ \int_{\Sigma^{l}_{h}} \alpha \cdot [W \cdot (x-\eta)]_n \, d\sigma - \int_{\Sigma^{i}_{h}} \alpha \cdot [W \cdot (x-\eta)]_n \, d\sigma. \qquad (3.23)$$

where $[W \cdot (x-\eta)]_n$ is the component of the vector $W(x) \cdot (x-y)$ normal to Σ and $\lambda > 4h$.

Let us divide the terms of the right-hand side of (3.23) into three groups and estimate each group separately.

Let us consider the first group. By virtue of Lemma 3 the function $W(x)$ on the surface Σ_λ^{l+}, Σ_λ^{l-} differs from $+1$, -1 respectively, by no more than

$$C_5\left(\frac{h}{\lambda}\right)^{C_6},$$

where C_j are constants dependent on M. Therefore

$$\int_{\Sigma_\lambda^{l+}} \alpha\cdot[W\cdot(x-\eta)]_n d\sigma - \int_{\Sigma_\lambda^{i+}} \alpha\cdot[W\cdot(x-\eta)]_n d\sigma + \int_{\Sigma_\lambda^{l-}} \alpha\cdot[W\cdot(x-\eta)]_n d\sigma$$

$$- \int_{\Sigma_\lambda^{i-}} \alpha\cdot[W\cdot(x-\eta)]_n d\sigma$$

$$\geq \mu(\Omega_\lambda^+) + \mu(\Omega_\lambda^-) - C_7\left(\frac{h}{\lambda}\right)^{C_6}. \tag{3.24}$$

It follows from (3.2) and the definition of $\Sigma_{h\,\lambda}^l$ that if the straight line defined by the angles φ, θ intersects $\Sigma_{h\,\lambda}^l$ then the following inequality holds

$$|f_1(\varphi,\theta)-f_2(\varphi,\theta)|\leq C_8\cdot\lambda. \tag{3.25}$$

It clearly follows from (3.25) that

$$\mu(\Omega)-\mu(\Omega_\lambda^+)-\mu(\Omega_\lambda^-)\leq C_9\cdot\lambda \tag{3.26}$$

where C_9 depends on M. Substituting (3.26) in (3.24) we obtain

$$\int_{\Sigma_\lambda^{l+}} \alpha\cdot[W\cdot(x-\eta)]_n d\sigma - \int_{\Sigma_\lambda^{i+}} \alpha\cdot[W\cdot(x-\eta)]_n d\sigma$$

$$+ \int_{\Sigma_\lambda^{l-}} \alpha\cdot[W\cdot(x-\eta)]_n d\sigma - \int_{\Sigma_\lambda^{i-}} \alpha\cdot[W\cdot(x-\eta)]_n d\sigma$$

$$\geq \mu(\Omega)-C_9\lambda-C_7\left(\frac{h}{\lambda}\right)^{C_6}. \tag{3.27}$$

We consider the second group. Clearly

$$\left|\int_{\Sigma_{h\,\lambda}^l} \alpha\cdot[W\cdot(x-\eta)]_n d\sigma - \int_{\Sigma_{h\,\lambda}^i} \alpha\cdot[W\cdot(x-\eta)]_n d\sigma\right|\leq \sigma(\Sigma_{h\,\lambda}^l)+\sigma(\Sigma_{h\,l}^i).$$

$$\tag{3.28}$$

We consider the function

$$\omega(\lambda)=\frac{\sigma(\Sigma^l)+\sigma(\Sigma^i)-\sigma(\Sigma_\lambda^l)-\sigma(\Sigma_\lambda^i)}{\sigma(\Sigma^l)+\sigma(\Sigma^i)}.$$

The function $\omega(\lambda)$ satisfies the conditions of Lemma 4, and thus for any $t < 1$ and n there will be a number k such that

$$\omega(t^{k+1}) - \omega(t^k) \leq \frac{1}{n}, \qquad k < n. \tag{3.29}$$

We set

$$h = t^{k+1}$$

$$\left(t \leq \frac{1}{4} \right) \tag{3.30}$$

$$\lambda = t^k$$

Substituting (3.30) and (3.29) in (3.28) we obtain

$$\left| \int\limits_{\Sigma_{h\lambda}^i} \alpha \cdot [W \cdot (x-\eta)]_n d\sigma - \int\limits_{\Sigma_{h\lambda}^i} \alpha \cdot [W \cdot (x-\eta)] \, d\sigma \right| \leq C_{10} \cdot \frac{1}{n}. \tag{3.31}$$

Finally we consider the third group. It follows from Lemma 3 that the function $W(x)$ on the surface Σ_h^i, Σ_h^i satisfies the inequality

$$|W(x)| \leq C_{11} \left(\frac{h}{\lambda} \right)^{C_{12}} = C_{11} t^{C_{12}}, \tag{3.32}$$

from which it follows that

$$\int\limits_{\Sigma_h^i} \alpha \cdot [W \cdot (x-\eta)]_n d\sigma - \int\limits_{\Sigma_h^i} \alpha \cdot [W \cdot (x-\eta)]_n d\sigma \leq C_{13} t^{C_{12}}. \tag{3.33}$$

Substituting (3.27), (3.31), (3.32) in (3.23) and substituting (3.22) and (3.23) in (3.19) we obtain

$$\mu(\Omega) \leq C_7 t^{C_6} + C_9 t^k + \frac{C_{10}}{n} + C_{13} t^{C_{12}} + \frac{C_4 \varepsilon_1}{t^{2k+2}}. \tag{3.34}$$

In (3.34) we set

$$t = \frac{1}{n} \tag{3.35}$$

where n satisfies the relation

$$\left(\frac{1}{n} \right)^{2n+1} \leq \varepsilon_1 \leq \left(\frac{1}{n-1} \right)^{2n-1}. \tag{3.36}$$

Substituting (3.35) and (3.36) in (3.34) we obtain

$$\mu(\Omega) \leq \frac{C}{n}$$

from which, by virtue of (3.2), the inequality (3.3) follows.

§ 2. A Class of Nonlinear Integral Equations

Let us consider the equation

$$A\varphi = f \tag{3.37}$$

where φ, f are continuous functions of x on the interval $[0, 1]$ and the operator A can be represented in the form

$$A\varphi = \int_0^1 \int_0^{\varphi(\xi)} P(x, \xi, \eta) \, d\eta \, d\xi \tag{3.38}$$

with $P(x, \xi, \eta)$ continuous and non-negative. The operator (3.38) is a special case of the URYSOHN operator

$$A\varphi = \int_0^1 K(x, \xi, \varphi(\xi)) \, d\xi$$

when the kernel $K(x, \xi, \eta)$ is a continuously differentiable function, satisfying the condition

$$K(x, \xi, 0) = 0$$

Let us formulate a uniqueness theorem for equation (3.37) for a class of operators A having the representation (3.38).

Theorem. Let the operator A satisfy the following conditions.

1. For any $\varphi_1, \varphi_2, \varphi_2(x) \geq \varphi_1(x)$ satisfying a LIPSCHITZ condition with some constant C, the operator $A_{\varphi_2}^{-1} A_{\varphi_1}$ exists where A_φ is the FRECHET derivative of the operator A at the point φ. Moreover, for any ψ satisfying a LIPSCHITZ condition with the constant C, the function $A_{\varphi_2}^{-1} A_{\varphi_1} \psi$ satisfies a LIPSCHITZ condition with the constant C_1 depending on C.

2. If $\varphi_2(x) > \varphi_1(x)$ the operator $A_{\varphi_2}^{-1} A_{\varphi_1}$ can be represented as

$$A_{\varphi_2}^{-1} A_{\varphi_1} \psi = \int_0^1 Q_{\varphi_2}[x, \xi, \varphi_2(\xi) - \varphi_1(\xi)] \psi(\xi) \, d\xi$$

where the function $Q_{\varphi_2}(x, \xi, \eta)$ is non-negative and continuous for $\eta > 0$.

3. The integral

$$\int_0^1 \int_0^1 Q_\varphi(x, \xi, \eta) \, d\xi \, d\eta$$

is convergent for any φ and x.

4. There exists a continuous function of two variables $\omega(t, h)$ dependent only on the constant C and satisfying the conditions,

a) $\omega(t, h) \to 0$, $h \to 0$, $t \neq 0$;

b) for any φ satisfying a LIPSCHITZ condition with the constant C the inequality

$$\frac{hQ_\varphi(x,\xi,h)}{W_\varphi(x,h)} \leq \omega(x-\xi,h)$$

holds where

$$W_\varphi(x,h) = \int\limits_0^h \int\limits_{x-h}^{x+h} Q_\varphi(x,\xi,h)\,d\xi d\eta\,.$$

Then the solution of equation (3.37) is unique within the class of functions satisfying a LIPSCHITZ condition.

It should be noted that this type of conditions is satisfied by the integral operators arising in the inverse problems of potential theory.

The classical example of an inverse problem is the one arising in the theory of NEWTONIAN potential formulated by NOVIKOV. Solving the planar inverse problem of NOVIKOV is equivalent to solving equation (3.37) with

$$P(x,\xi,h) = \ln\left[r^2 + h^2 - 2r\eta\cos 2\pi(x-\xi)\cdot\eta\right], \quad 0 < \varphi < r\,.$$

If $\varphi_1(x)$, $\varphi_2(x)$, and $\psi(x)$ are continuous functions, $\varphi_2(x) > \varphi_1(x)$, then in this case the function $A_{\varphi_2}^{-1} A_{\varphi_1}\psi$ is a density of a single layer potential with weight $\varphi_2(x)$ distributed on the curve defined by the polar equation

$$\varrho = \varphi_2(x) \qquad (\Gamma_2)$$

($2\pi x$ is the polar angle) and satisfying the following condition, the logarithmic potential of a single layer with density $\varphi_1(x)\psi(x)$ on the curve

$$\varrho = \varphi_1(x) \qquad (\Gamma_1)$$

coincides with the single layer potential with density $A_{\varphi_2}^{-1} A_{\varphi_1}\psi$ outside the domain bounded by Γ_2.

It is easy to show that in this case the operator $A_{\varphi_2}^{-1} A_{\varphi_1}$ satisfies the conditions 2, 3, 4 of the theorem.

Let us introduce some notation. Let φ_j be continuous functions and let

$$u(x,\varphi_1,\varphi_2) = \max \varphi_j(x),$$

$$j = 1,2$$

$$v(x,\varphi_1,\varphi_2) = \min \varphi_j(x),$$

$$W(x,\varphi_1,\varphi_2,\varphi_3) = \min_{i,j=1,2,3} u(x,\varphi_i,\varphi_j) = \max_{i,j=1,2,3} v(x,\varphi_i,\varphi_j)\,.$$

Let α_j $(j = 1, \ldots)$ be sets of points of the interval $[0, 1]$. We denote by $\varrho(\alpha_1,\alpha_2)$ the distance between the sets α_1, α_2.

We denote by $\alpha(\varphi_1,\varphi_2)$ the set of values of x for which

$$\varphi_1(x) > \varphi_2(x)$$

and by $\bar{\alpha}(\varphi_1,\varphi_2)$ the set of values of x for which

$$\varphi_1(x) = \varphi_2(x)\,.$$

Let λ_j $(j = 1, ...)$ be some real numbers. We denote by $\alpha\,(\varphi_1, \varphi_2, \lambda)$ the set defined by the relations

$$x \in \alpha\,(\varphi_1, \varphi_2, \lambda)$$

if

$$x \in \alpha\,(\varphi_1, \varphi_2), \qquad \varrho\,[x, \bar{\alpha}\,(\varphi_1, \varphi_2)] \geq \lambda$$

and by $\bar{\alpha}\,(\varphi_1, \varphi_2, \lambda)$ the set

$$\bar{\alpha}\,(\varphi_1, \varphi_2, \lambda) = \alpha\,(\varphi_1, \varphi_2) \oplus \bar{\alpha}\,(\varphi_1, \varphi_2) \ominus \alpha\,(\varphi_1, \varphi_2, \lambda)$$

while by $\bar{\alpha}\,(\varphi_1, \varphi_2, \lambda_1, \lambda_2)$ the set

$$\bar{\alpha}\,(\varphi_1, \varphi_2, \lambda_1, \lambda_2) = \bar{\alpha}\,(\varphi_1, \varphi_2, \lambda_1) \ominus \bar{\alpha}\,(\varphi_1, \varphi_2, \lambda_2).$$

The proof of the theorem is based on the following lemma.

Lemma. Let A be an operator satisfying the conditions of the theorem and let φ_1 and φ_2 be functions satisfying a LIPSCHITZ condition with the constant C, and such that the sets $\alpha\,(\varphi_1, \varphi_2)$ are not empty. Then for any sufficiently small positive λ and h there exist functions $\bar{\varphi}_1, \bar{\varphi}_2, \psi$ possessing the following properties,

1. $u\,(x, \varphi_1, \varphi_2) \geq \bar{\varphi}_j(x) \geq v\,(x, \varphi_1, \varphi_2), \quad j = 1, 2;$

2. $A\varphi_2 - A\varphi_1 = A\bar{\varphi}_2 - A\bar{\varphi}_1 + A_{\bar{u}}\psi, \quad \bar{u} = u\,(x, \bar{\varphi}_1, \bar{\varphi}_2);$

3. $\psi(x) \geq 0, \quad x \in \alpha\,(\bar{\varphi}_2, \bar{\varphi}_1);$

    $\psi(x) \leq 0, \quad x \in \alpha\,(\bar{\varphi}_1, \bar{\varphi}_2);$

4. $\varrho\,[\alpha\,(\bar{\varphi}_1, \bar{\varphi}_2), \; \alpha\,(\bar{\varphi}_2, \bar{\varphi}_1)] \geq \lambda;$

5. $\max\limits_{x} |\bar{\varphi}_2(x) - \bar{\varphi}_1(x)| = h;$

6. the functions $\psi, \bar{\varphi}_j$ $(j = 1, 2)$ satisfy a LIPSCHITZ condition with the constant C_1 depending on C.

We prove the above lemma. Let φ_1, φ_2 be functions satisfying a LIPSCHITZ condition with the constant C, let

$$A\varphi_1 - A\varphi_2 = f$$

and let λ, h be some sufficiently small constants.

We denote by $\gamma_{m\lambda}$ the function defined by the relations

a) $\gamma_{m\lambda} = \dfrac{1}{m}(\varphi_2 - \varphi_1), \quad x \in \bar{\alpha}\,(\varphi_2, \varphi_1, \lambda);$

b) $\gamma_{m\lambda} = \dfrac{1}{m}(\varphi_1 - \varphi_2), \quad x \in \bar{\alpha}\,(\varphi_1, \varphi_2, \lambda);$

$\gamma_{m\lambda} = 0, \quad x \in \alpha\,(\varphi_1, \varphi_2, 2\lambda) \oplus \alpha\,(\varphi_2, \varphi_1, 2\lambda);$

c) $\gamma_{m\lambda}$ is a continuous function on the interval $[0, 1]$ and it changes linearly on the set $\bar{\alpha}\,(\varphi_1, \varphi_2, 2\,\lambda, \lambda)$.

We consider the sequences of a function $\varphi_{1\,mk\lambda}, \varphi_{2\,mk\lambda}, \psi_{mk\lambda}\;(k = 1, \ldots)_y$

$$\left.\begin{array}{l} \varphi_{jmo\lambda} = \varphi_j \\[4pt] \varphi_{jmk+1\lambda} = \varphi_{jmk\lambda} + \delta_{jmk\lambda} \end{array}\right\} \quad (j=1,2)$$

$$\psi_{mo\lambda} = 0, \qquad \psi_{m,\,k+1,\,\lambda} = \psi_{mk\lambda} + \tilde{\delta}_{mk\lambda},$$

where

a) $\delta_{1mk\lambda} = v\,(\gamma_{m\lambda}, \varphi_{2mk\lambda} - \varphi_{1mk\lambda}), \qquad x \in \alpha\,(\varphi_2, \varphi_1),$

   $\delta_{2mk\lambda} = v\,(\gamma_{m\lambda}, \varphi_{1mk\lambda} - \varphi_{2mk\lambda}), \qquad x \in \alpha\,(\varphi_1, \varphi_2);$

b) $\delta_{1mk\lambda} = W\,[A_{u_{mk\lambda}}^{-1} A_{v_{mk\lambda}} \tilde{\delta}_{mk\lambda};\; 0;\; \varphi_{2mk\lambda} - \varphi_{1mk\lambda}], \quad x \in \alpha\,(\varphi_1, \varphi_2),$

   $\delta_{2mk\lambda} = W\,[A_{u_{mk\lambda}}^{-1} A_{v_{mk\lambda}} \tilde{\delta}_{mk\lambda};\; 0;\; \varphi_{1mk\lambda} - \varphi_{2mk\lambda}], \quad x \in \alpha\,(\varphi_2, \varphi_1),$

   $u_{mk\lambda}(x) = u\,(x, \varphi_{1mk\lambda}, \varphi_{2mk\lambda}),$

   $v_{mk\lambda}(x) = v\,(x, \varphi_{1mk\lambda}, \varphi_{2mk\lambda}),$

   $\tilde{\delta}_{mk\lambda}(x) = \delta_{1mk\lambda}(x), \qquad\qquad\qquad x \in \alpha\,(\varphi_2, \varphi_1),$

   $\tilde{\delta}_{mk\lambda}(x) = \delta_{2mk\lambda}(x), \qquad\qquad\qquad x \in \alpha\,(\varphi_1, \varphi_2);$

c) $\delta_{jmk\lambda} = 0, \qquad\qquad\qquad\qquad\qquad x \in \bar{\alpha}\,(\varphi_1, \varphi_2);$

d) $\tilde{\delta}_{mk\lambda} = A_{u_{mk\lambda}}^{-1} A_{v_{mk\lambda}} \tilde{\delta}_{mk\lambda} - \delta_{1mk\lambda}, \qquad x \in \alpha\,(\varphi_1, \varphi_2),$

   $\tilde{\delta}_{mk\lambda} = A_{u_{mk\lambda}}^{-1} A_{v_{mk\lambda}} \tilde{\delta}_{mk\lambda} - \delta_{2mk\lambda}, \qquad x \in \alpha\,(\varphi_2, \varphi_1),$

   $\tilde{\delta}_{mk\lambda} = A_{u_{mk\lambda}}^{-1} A_{v_{mk\lambda}} \tilde{\delta}_{mk\lambda}, \qquad\qquad x \in \bar{\alpha}\,(\varphi_1, \varphi_2).$

One can easily see that

$$|A\varphi_{2mk\lambda} - A\varphi_{1mk\lambda} + A_{u_{mk\lambda}} \psi_{mk\lambda} - f| \le \frac{C_1 k}{m} \mu\left(\frac{1}{m}\right) \qquad (3.39)$$

where C_1 is a constant depending on C and $\mu\,(t)$ is the modulus of continuity of the function $P\,(x, \xi, \eta)$.

Indeed by virtue of the definition of the FRECHET derivative

$$\|A\varphi_{jmk+1\lambda} - A\varphi_{jmk\lambda} - A_{\varphi_{jmk\lambda}} \delta_{jmk\lambda}\| = 0\;(\|\delta_{jmk\lambda}\|). \qquad (3.40)$$

From the representation (3.38) of the operator A it follows that the right-hand side of (3.40) satisfies the inequality

$$0(\|\delta_{jmk\lambda}\|) \le \frac{C_2}{m} \mu\left(\frac{1}{m}\right).$$

The inequality (3.39) now follows by virtue of the definitions of the functions $\varphi_{jmk\lambda}, \delta_{jmk\lambda}, \psi_{mk\lambda}.$

Next, from the conditions imposed on the operator $A_{\varphi_2}^{-1} A_{\varphi_2}$ it follows that for $k \leq m$ the functions $\varphi_{jmk\lambda}, \psi_{mk\lambda}$ satisfy a LIPSCHITZ condition with a constant C_3 dependent only on C and that for $k \geq m$

$$
\begin{aligned}
\varphi_{jm, k+1, \lambda} &= \varphi_{jmk\lambda} = \tilde{\varphi}_{jm\lambda}, \\
\psi_{m, k+1, \lambda} &= \psi_{mk\lambda} = \check{\psi}_{m\lambda}.
\end{aligned}
\tag{3.41}
$$

The functions $\tilde{\varphi}_{jm\lambda}$ in (3.41) evidently satisfy the conditions

$$
\begin{aligned}
&\varrho\left[\alpha(\tilde{\varphi}_{1m\lambda}, \tilde{\varphi}_{2m\lambda}), \alpha(\tilde{\varphi}_{2m\lambda}, \tilde{\varphi}_{1m\lambda})\right] \geq \lambda, \\
&v(\varphi_1, \varphi_2) \leq \tilde{\varphi}_{jm\lambda} \leq u(\varphi_1, \varphi_2)
\end{aligned}
\tag{3.42}
$$

and for any sufficiently small h, λ

$$
\max_x |\tilde{\varphi}_{2m\lambda} - \tilde{\varphi}_{1m\lambda}| > 2h.
\tag{3.43}
$$

We consider now the sequences of functions $\tilde{\varphi}_{1m\lambda}, \tilde{\varphi}_{2m\lambda}, \check{\psi}_{m\lambda}$ $(m = 1, \ldots)$. By virtue of the fact that all the elements of sequences are bounded and satisfy a LIPSCHITZ condition with the constant C_1 independent of m, one may choose from the sequences uniformly convergent subsequences

$$
\tilde{\varphi}_{1m_q\lambda}, \tilde{\varphi}_{2m_q\lambda} \quad \check{\psi}_{m_q\lambda} \qquad (q = 1, \ldots).
$$

By virtue of (3.42) and (3.43) the functions

$$
\begin{aligned}
\tilde{\varphi}_j &= \lim_{q \to \infty} \tilde{\varphi}_{jm_q\lambda}, \qquad (j = 1, 2) \\
\check{\psi} &= \lim_{q \to \infty} \check{\psi}_{m_q\lambda}
\end{aligned}
$$

satisfy the following relations

$$
\begin{aligned}
&\varrho\left[\alpha(\tilde{\varphi}_1, \tilde{\varphi}_2), \alpha(\tilde{\varphi}_2, \tilde{\varphi}_1)\right] \geq \lambda, \\
&A\tilde{\varphi}_1 - A\tilde{\varphi}_2 + A_{\check{u}}\check{\psi} = f, \\
&v(\varphi_1, \varphi_2) \leq \tilde{\varphi}_j \leq u(\varphi_1, \varphi_2), \\
&\max_x |\tilde{\varphi}_2 - \tilde{\varphi}_1| > h
\end{aligned}
\tag{3.44}
$$

and a LIPSCHITZ condition with constant C_4.

We denote by γ_m the function defined by the relations

$$
\begin{aligned}
\gamma_m &= \frac{1}{m}(\tilde{\varphi}_2 - \tilde{\varphi}_1), && x \in \alpha(\tilde{\varphi}_2, \tilde{\varphi}_1), \\[1.5ex]
\gamma_m &= \frac{1}{m}(\tilde{\varphi}_1 - \tilde{\varphi}_2), && x \in \alpha(\tilde{\varphi}_1, \tilde{\varphi}_2) \\[1.5ex]
\gamma_m &= 0 && x \in \bar{\alpha}(\tilde{\varphi}_1, \tilde{\varphi}_2)
\end{aligned}
$$

and consider the sequence of functions $\varphi_{1mk}, \varphi_{2mk}, \psi_{mk}$ $(k = 1, \ldots)$

$$\varphi_{jmo} = \tilde{\varphi}_j \qquad (j = 1, 2),$$

$$\varphi_{jm, k+1} = \varphi_{jmk} + \delta_{jmk},$$

$$\psi_{mo} = \tilde{\psi},$$

$$\psi_{m, k+1} = \psi_{mk} + \tilde{\delta}_{mk},$$

where $\delta_{jmk}, \tilde{\delta}_{mk}$ are functions defined by relations similar to relations a), b), c) defining the functions $\delta_{jmk\lambda}, \tilde{\delta}_{mk\lambda}$.

In complete analogy with (3.39) we obtain

$$|A\varphi_{2mk} - A\varphi_{1mk} + A_{u_{mk}}\psi_{mk} - f| \leq \frac{C_5 k}{m} \mu\left(\frac{1}{m}\right), \qquad (3.45)$$

$$\max |\varphi_{2mp} - \varphi_{1mp}| > h,$$

$$\max |\varphi_{2mp+1} - \varphi_{1mp+1}| \leq h. \qquad (3.46)$$

By virtue of the definition of the sequences φ_{jmk} we can obviously find a number p, $0 \leq p \leq m$, satisfying (3.46).

We denote by $\bar{\varphi}_{jm}, \bar{\psi}_m$ the functions $\bar{\varphi}_{jm} = \varphi_{jmp}, \bar{\psi}_m = \bar{\psi}_{mp}$ and consider the sequences

$$\bar{\varphi}_{jm}, \bar{\psi}_m \qquad (m = 1, \ldots).$$

In analogy with the sequences $\varphi_{jm\lambda}, \psi_{m\lambda}$ $(m = 1, \ldots)$ we may select uniformly convergent sub-sequences $\bar{\varphi}_{jm_q}, \bar{\psi}_{m_q}$ $(q = 1, \ldots)$ from the sequences $\bar{\varphi}_{jm}, \bar{\psi}_m$. By virtue of the above, the functions

$$\bar{\varphi}_j = \lim_{q \to \infty} \bar{\varphi}_{jm_q} \qquad (j = 1, 2),$$

$$\bar{\psi} = \lim_{q \to \infty} \bar{\psi}_{m_q}$$

satisfy the relations 1)−6) and the lemma is proved.

We now prove the theorem. We suppose the solution of (3.37) to be non-unique within the class of Lipschitz continuous functions, i.e., there exist distinct functions φ_1, φ_2 satisfying a Lipschitz condition with the constant C and

$$A\varphi_1 - A\varphi_2 = 0. \qquad (3.47)$$

It follows from the Lemma that for any sufficiently small h and λ there exist functions $\bar{\varphi}_1, \bar{\varphi}_2, \bar{\psi}$ satisfying the conditions 1)−6) of the Lemma and therefore, according to (3.47),

$$A\bar{\varphi}_1 - A\bar{\varphi}_2 + A_u\bar{\psi} = 0.$$

Applying the operator $A_{\bar{u}}^{-1}$ to (3.48) and making use of the conditions 2) and 3) of the theorem, we obtain

$$\int\limits_0^1 \int\limits_0^{\bar u\,(\xi)-v\,(\xi)} Q_{\bar u}[x,\xi,\eta]\,d\eta\cdot \text{sign}\,[\bar\varphi_2(\xi)-\bar\varphi_1(\xi)]\,d\xi+\bar\Psi(x)=0. \quad (3.49)$$

We denote by $\bar x$ a value of the independent variable x such that

$$|\bar\varphi_2(\bar x)-\bar\varphi_1(\bar x)|=\max_x |\bar\varphi_2(x)-\bar\varphi_1(x)|.$$

Because of the properties of $\bar\varphi_j$

$$|\bar\varphi_2(\bar x)-\bar\varphi_1(\bar x)|=h. \quad (3.50)$$

Let, for the sake of definiteness,

$$\bar\varphi_2(\bar x)>\bar\varphi_1(\bar x),$$

i.e.,

$$\bar\varphi_2(\bar x)-\bar\varphi_1(\bar x)=h. \quad (3.51)$$

We put $x = \bar x$ in (3.49) and divide the integral in the left-hand side of (3.49) into three parts

$$\int\limits_{\bar x-h_1}^{\bar x+h_1} \int\limits_0^{\bar u\,(\xi)-v\,(\xi)} Q_{\bar u}(\bar x,\xi,\eta)\,d\eta\cdot \text{sign}\,[\bar\varphi_2(\xi)-\bar\varphi_1(\xi)]\,d\xi$$

$$+\int\limits_0^{\bar x-h_1} \int\limits_0^{\bar u\,(\xi)-v\,(\xi)} Q_{\bar u}(\bar x,\xi,h)\,d\eta\cdot \text{sign}\,[\bar\varphi_2-\bar\varphi_1]\,d\xi \quad (3.52)$$

$$+\int\limits_{\bar x+h_1}^1 \int\limits_0^{\bar u\,(\xi)-v\,(\xi)} Q_{\bar u}(\bar x,\xi,h)\,d\eta\cdot \text{sign}\,[\bar\varphi_2-\bar\varphi_1]\,d\xi+\bar\Psi(\bar x)=0,$$

where

$$h_1=\frac{h}{4\,C_1}.$$

We denote the integrals on the left side of (3.52) by I_1, I_2 and I_3 respectively and estimate them separately. Let us consider the first integral. From (3.51) and condition 6) of the lemma it follows that

$$|\bar\varphi_2(x)-\bar\varphi_1(x)|\ge\frac{h}{2} \qquad |x-\bar x|\ge\frac{h}{4\,C_1}, \quad (3.53)$$

from which

$$I_1\ge W_{\bar u}(\bar x,h),$$

where $W_{\bar u}(x, h)$ is the function defined in condition 4) of the theorem.

We consider now the second and third integrals. From condition 4) of the lemma we have,

$$I_2\ge \int\limits_0^{\bar x-\lambda} \int\limits_0^{\bar u\,(\xi)-v\,(\xi)} Q_{\bar u}(\bar x,\xi,\eta)\,d\eta\cdot \text{sign}\,[\bar\varphi_2-\bar\varphi_1]\,d\xi, \quad (3.54)$$

$$I_3\ge \int\limits_{\bar x+\lambda}^1 \int\limits_0^{\bar u\,(\xi)-v\,(\xi)} Q_{\bar u}(\bar x,\xi,\eta)\,d\eta\cdot \text{sign}\,[\bar\varphi_2-\bar\varphi_1]\,d\xi,$$

while from (3.54), condition 4) of the theorem and condition 5) of the lemma we obtain

$$I_2 + I_3 \geq -\omega(\lambda, h) \cdot W_{ii}(\bar{x}, h). \tag{3.55}$$

From condition 3) of the lemma and from (3.51) it follows that the last term in the left-hand side of (3.52) is non-negative, i.e.,

$$\bar{\psi}(\bar{x}) \geq 0.$$

From (3.54), (3.55) and (3.56) it follows that for sufficiently small h there holds the inequality

$$I_1 + I_2 + I_3 + \bar{\psi}(\bar{x}) > 0$$

contradicting the equality (3.52). Thus the theorem is proved.

§ 3. Inverse Problems for Some Non-Newtonian Potentials

Inverse problems for NEWTONIAN potentials have unique solutions only under very essential restrictions on the nature of the density distribution. An example constructed by P. S. NOVIKOV indicates that the solution of the inverse problem of NEWTONIAN potential may be non-unique even in the class of densities which are piecewise constant with a given constant and are contained in simply connected domains. An analogous situation seems to take place also for potentials corresponding to the fundamental solutions of other elliptic equations.

As will be shown in this section, for inverse problems with some Non-NEWTONIAN potentials not connected with differential equations, uniqueness holds for much milder restrictions on the nature of the density; its continuity and finiteness will be quite sufficient.

Let x, ξ be n-dimensional vectors with components $(x_1, \ldots, x_n)$, $(\xi_1, \ldots, \xi_n)$, let D, D_1 be non-intersecting simply connected bounded domains of n-dimensional space and let

$$u(x) = \int_D \frac{1}{r^p(x, \xi)} \cdot \varrho(\xi) d\xi, \tag{3.56}$$

where

$$r(x, \xi) = |x - \xi|.$$

Given the function $u(x)$ in D_1, it is required to reconstruct the potential density ϱ in D.

Theorem. For $p > n + 2$ the solution of the problem is unique in the class of continuous $\varrho(x)$, i.e., the continuous function $\varrho(x)$ is uniquely determined by the values of the function $u(x)$ in the domain D_1.

For our proof we shall examine the function

$$v(x) = \int_D \frac{1}{R^p(x, y, \xi)} \varrho(\xi) \, d\xi \qquad (3.57)$$

where

$$R(x, y, \xi) = \sqrt{r^2(x, \xi) + |y|^2}$$

and y is a vector with components $(y_1, \ldots, y_{m+1})$, $m = p - n - 3$.

The function $v(x, y)$ will clearly be a harmonic function of the variables x, y with cylindrical symmetry. From the results of § 6, Chapter II it follows that the function $v(x, y)$ is uniquely determined on a set Q of x, y space:

$$x, y \in Q, \text{ if } x \bar{\in} D \text{ or } y \neq 0;$$

by its values on the set

$$x \in D_1, \qquad y = 0.$$

The function $v(x, y)$ is a NEWTONIAN potential of a simple layer with density $\varrho(x)$ distributed on the n-dimensional manifold $x \in D$, $y = 0$ in the $(n + m + 1)$-dimensional x, y space. It is clear that $\varrho(x)$ is uniquely determined from $v(x, y)$. Hence the theorem is proved.

We note that the method used for the proof of the uniqueness theorem permits one to obtain estimates characterizing the stability of the solution.

§ 4. An Inverse Problem for the Wave Equation

In this section we shall prove a theorem giving the uniqueness for the solution of an inverse problem for the wave equation.

We consider the equation

$$n^2 \frac{\partial^2 u}{\partial t^2} = \Delta u \qquad (3.58)$$

where u is a function of the three variables x, y, t and n is a function of the variables x, y.

We consider the following problem: the domain D_0 is given in the x, y plane. The function $n(x, y) > 0$ is continuous and identically equal to one outside of D_0. Moreover, in some domain D_1 with $D_1 \cap D_0$ empty, there is given a family G of solutions of (3.58) for all $t > 0$. It is required to determine the function $n(x, y)$ inside D_0.

We note that a similar problem in the case of one variable was considered in the paper [59]; quite close to the problem being considered there, is the well-known STURM-LIOUVILLE inverse problem.

We formulate a uniqueness theorem for the above problem for a certain family G.

Theorem. Let D_0 and D_1 be bounded and simply connected domains, and D_2 a bounded, simply connected domain, which does not intersect with D_0 or D_1. Let G be a set of solutions of (3.58) satisfying the following initial conditions

$$u(x, y, 0) = 0;$$

$$\frac{\partial}{\partial t} u(x, y, 0) = \delta(x - x_0, y - y_0) \tag{3.59}$$

where $Q(x_0, y_0)$ is any point of D_2.

Then the solution of the inverse problem is unique, i.e., the function $n(x, y)$ is uniquely determined within D_0.

We give a brief sketch of the proof. We denote the solution of the CAUCHY problem (3.59) for equation (3.58) by $u(x, y, x_0, y_0, t)$ and consider the function

$$v(x, y, x_0, y_0 \cdot \lambda) = \int_0^\infty u(x, y, x_0, y_0, t) \cos \lambda t \, dt .$$

One can easily see that

$$\Delta v = -\delta(x - x_0, y - y_0) - \lambda^2 n^2 v . \tag{3.60}$$

The function v is the fundamental solution of the HELMHOLTZ equation

$$\Delta v = -\lambda^2 n^2 v$$

with the singularity at the point $Q(x_0, y_0)$. It is well-known that a fundamental solution for an elliptic equation with analytic coefficients is an analytic function of both the independent variables x and y and the coordinates of the singularity x_0, y_0 everywhere away from the singularity.

According to the conditions of the theorem the function v is given in the domain D_{12} of the 4-dimensional space $R(x, y, x_0, y_0)$, the direct product of the domains D_1, D_2 of the spaces $P(x, y)$ and $Q(x_0, y_0)$. The function $n(x, y)$ is identically equal to one outside of D_0. Hence, by virtue of the uniqueness of the analytical continuation, the function v may be considered to be given everywhere outside of the domain $D_{00} \equiv D_0 \times D_0$.

Let D_3 be a bounded domain containing the domains D_0, D_1 and D_2 and let

$$r^2 = (x - \xi)^2 + (y - \eta)^2, \quad r_0^2 = (x_0 - \xi)^2 + (y_0 - \eta)^2 .$$

From (3.60) it follows that for $P \in D_3$

$$v(x, y, x_0, y_0, \lambda) = \frac{1}{2\pi} \ln\left[(x - x_0)^2 + (y - y_0)^2\right]$$

$$- \frac{1}{2\pi} \lambda n^2(x, y) \int_{D_0} v(\xi, \eta, x_0, y_0 \cdot \lambda) \ln r \, d\xi \, d\eta$$

$$+ \tilde{v}(x, y, x_0, y_0, \xi); \tag{3.61}$$

$$\tilde{v} = \frac{1}{2\pi} \int_{\Gamma_3} \left(v \frac{\partial}{\partial n} \ln r - \frac{\partial v}{\partial n} \ln r \right) ds,$$

where Γ_3 is the boundary of D_3.

We denote by $v_1(x, y, x_0, y_0)$ the function

$$v_1 = \frac{\partial}{\partial \lambda} \left[\frac{\partial^2 v(x, y, x_0, y_0, 0)}{\partial x \cdot \partial x_0} + \frac{\partial^2 v(x, y, x_0, y_0, 0)}{\partial y \cdot \partial y_0} \right].$$

It can be easily shown that the function v_1, for $R \in D_{33} \equiv D_3 \times D_3$, equals

$$v_1 = \int_{D_0} n(\xi, \eta) \frac{(x - \xi)(x_0 - \xi) + (y - \eta)(y_0 - \eta)}{r^2 \cdot r_0^2} d\xi \cdot d\eta + \tilde{v}_1 \qquad (3.62)$$

where

$$\tilde{v}_1(x, y, x_0, y_0) = \frac{\partial}{\partial \lambda} \left[\frac{\partial^2 \tilde{v}}{\partial x \cdot \partial x_0} + \frac{\partial^2 \tilde{v}}{\partial y \cdot \partial y_0} \right].$$

By virtue of (3.62) the functions $v_1, \tilde{v}_1$ for $R \in D_{33}$ and $R \in D_{00}$ are analytic functions of the variables x, y, x_0, y_0.

By setting

$$x = x_0, \quad y = y_0; \quad P(x, y) \in D_1$$

in (3.62) we obtain

$$v_2 = (v_1(x, y, x, y) - \tilde{v}_1(x, y, x, y) = \int_{D_0} n(\xi, \eta) \cdot \frac{1}{r^2} d\xi d\eta. \qquad (3.63)$$

The assertion of the theorem follows immediately from the results of § 3 and (3.63).

We note that the theorem generalizes to the case of the wave equation in a space of an arbitrary number of dimensions as well as to the heat equation and some equations of hyperbolic and elliptic type of higher orders.

§ 5. On a Class of Inverse Problems for Differential Equations

In this section a statement more general than that of the previous section is considered. A class of inverse problems is reduced to linear integral equations of the first kind which are then investigated by methods differing from those of the previous section.

We consider the equation

$$P_1\left(\frac{\partial}{\partial x_j}\right) u(x, y) = P_3\left(\frac{\partial}{\partial y_j}\right) P_2\left(\frac{\partial}{\partial x_j}\right) u(x, y), \qquad (3.64)$$

where x, y are vectors with components $(x_1, ..., x_n)$, $(y_1, ..., y_n)$ and P_j are polynomials with coefficients continuously depending on x.

In regard to P_1, we assume that the equation

$$P_1\left(\frac{\partial}{\partial x_j}\right)v(x)=0$$

has a fundamental solution $G(x, x^0)$ defined in the entire x, x_0 space*.

We pose the inverse problem for equation (3.64) which will be considered in this section.

Let D_0, D_1 be some bounded domains of x space, where the intersection $D_1 \cap D_0$ is empty. The coefficients of the polynomials P_1, P_2 are given for the entire x space and the coefficients of the polynomial P_3 are given everywhere outside of D_0. In addition, let a family of solutions $u(x, y, \xi)$ of (3.64) depending on the parameter $\xi(\xi_1, ..., \xi_p)$, be given in the domain D_1.

The functions $u(x, y, \xi)$ satisfy the conditions

$$\frac{\partial^r}{\partial y_1^{k_1} ... \partial y_m^{k_m}}u(x,0,\xi)=f_{k_1 \cdots k_m}(x,\xi),$$

$$\lim_{\varrho \to \infty} \int_{\Sigma_{\varrho x_0}} \left|\frac{\partial^{r_1}u(x,0,\xi)}{\partial x_1^{i_1} ... \partial x_n^{i_n}}\right| \cdot \left|\frac{\partial^{r_2}G(x,x^0)}{\partial x_1^{j_1} ... \partial x_n^{j_n}}\right| d\sigma_x=0, \qquad (3.65)$$

$$\left|\frac{\partial^r u(x,y,\xi)}{\partial y_1^{k_1} ... \partial y_m^{k_m}}\right| \leq C_1 + |y|^{C_2}$$

where the indices $k_1, ..., k_m$ run over the values which are less than the maximum values of the corresponding indices in the polynomial

$$P_3\left(\frac{\partial}{\partial y_j}\right)$$

and $i_1 + j_1, ..., i_n + j_n$ run over a similar set with respect to the polynomial

$$P_1\left(\frac{\partial}{\partial x_j}\right).$$

$\Sigma_{\varrho x_0}$ is a sphere of radius ϱ with the center at the point x^0, $f(x, \xi)$ are given finite functions, and C_1, C_2 are some constants.

It is required to determine the coefficients of the polynomial P_3 in the domain D_0.

In the case when $n = m = 1$ the above inverse problem is equivalent to the known STURM-LIOUVILLE inverse problem for ordinary differential equations.

---

* The existence of a fundamental solution has been proved, say, for elliptic equations of the second order and also for ordinary regular equations with constant coefficients.

We denote by $v(x, \lambda, \xi)$ the function

$$v(x, \lambda, \xi) = \int_0^\infty u(x, y, \xi) \exp\{-(\lambda, y)\} \, dy$$

(λ is a vector with components $\lambda_1, \ldots, \lambda_m$).

By virtue of (3.64) and (3.65) the function $v(x, \lambda, \xi)$ satisfies the following differential equation

$$P_1\left(\frac{\partial}{\partial x_j}\right) v = P_3(-\lambda_j) \cdot P_2\left(\frac{\partial}{\partial x_j}\right) v + \psi, \qquad (3.66)$$

where

$$\psi(x, \lambda, \xi) = Q\left(\lambda_j, \frac{\partial}{\partial x_j}, \frac{\partial}{\partial y_k}\right) \cdot u(x, 0, \xi)$$

is a function which appears as a result of integrating by parts the expression for

$$P_3\left(\frac{\partial}{\partial y_j}\right) \cdot v$$

and Q is a polynomial.

We multiply both sides of (3.66) by the function $G(x, x^0)$ and integrate over the whole x space. From inequalities (3.65) and (3.66) it follows that the integrals will converge, and by integrating the corresponding expressions by parts, the boundary terms at infinity will be equal to zero.

Thus, we find that the function $v(x^0, \lambda, \xi)$ satisfies the integral relation

$$v(x^0, \lambda, \xi) = \int G(x, x^0) P_3(\lambda_j) P_2\left(\frac{\partial}{\partial x_j}\right) v \, dx + \int G(x, x^0) \psi \, dx. \qquad (3.67)$$

From (3.67) it follows that the function $v(x, \lambda, \xi)$ is an analytic function of λ in some neighborhood of the origin. We consider

$$v_{k_1 \cdots k_m}(x, \xi) = \frac{\partial^r}{\partial \lambda_1^{k_1} \ldots \partial \lambda_m^{k_m}} v(x, 0, \xi),$$

$$\psi_{k_1 \cdots k_m}(x, \xi) = \frac{\partial^r}{\partial \lambda_1^{k_1} \ldots \partial \lambda_m^{k_m}} \psi(x, 0, \xi),$$

$$P_{k_1 \cdots k_m} = \frac{\partial^r}{\partial \lambda_1^{k_1} \ldots \partial \lambda_m^{k_m}} P_3(0).$$

By virtue of (3.67) the functions $v_{k_1 \cdots k_m}, \psi_{k_1 \cdots k_m}, P_{k_1 \cdots k_m}$ satisfy the following system of relations

$$v_{k_1 \cdots k_m}(x^0, \xi) = \int G(x, x^0)\left\{\sum_{j_q + i_q = k_q} P_{j_1 \cdots j_m} P_2\left(\frac{\partial}{\partial x_j}\right) \times v_{i_1 \cdots i_m}(x, \xi)\right\} dx$$

$$+ \int G(x, x^0) \psi_{k_1 \cdots k_m}(x, \xi) \, dx, \qquad (3.68)$$

$$v_{0 \cdots 0}(x^0, \xi) = \int G(x, x^0) \psi_{0 \cdots 0}(x, \xi) \, dx.$$

Relations (3.68) may be regarded as a recursive system of linear integral equations of the first kind in D_0 for the coefficients $P_{j_1 \ldots j_m}$ of the polynomial P_3. In fact it follows from (3.68) that the function $v_{0 \ldots 0}(x^0, \xi)$ may be regarded as given in the whole space.

We consider now the function

$$v_{\underbrace{0 \ldots 1 \ldots 0}_{q}}(x^0, \xi) = \int_{D_0} G(x, x^0) P_{\underbrace{0 \ldots 1 \ldots 0}_{q}} P_2\left(\frac{\partial}{\partial x_j}\right) v_{0 \ldots 0}(x, \xi)\, dx$$

$$+ \int_{D_0} G(x, x^0) P_{\underbrace{0 \ldots 1 \ldots 0}_{q}} P_2\left(\frac{\partial}{\partial x_j}\right) v_{\ldots 0}(x, \xi)\, dx$$

$$+ \int G(x, x^0) \psi_{\underbrace{0 \ldots 1 \ldots 0}_{q}}(x, \xi)\, dx. \tag{3.69}$$

If $x^0 \in D_1$ then in the equality (3.69) all the terms may be regarded as given with the exception of the first one in the right-hand side, and therefore (3.69) may be regarded as a linear integral equation of the first kind for the function

$$P_{\underbrace{0 \ldots 1 \ldots 0}_{q}}(x), \quad x \in D_0$$

with the kernel

$$K(x, x^0, \xi) = G(x, x^0) P_2\left(\frac{\partial}{\partial x_j}\right) v_{0 \ldots 0}(x, \xi), \quad x^0 \in D_1.$$

If the above equation has a unique solution for any q then the equations defining the function

$$P_{\underbrace{0 \ldots 1}_{q}\ \underbrace{0 \ldots 1 \ldots 0}_{q_1}}(x)$$

is found in a similar way, and so on.

We will now carry out a detailed consideration of the system (3.68) in the case in which the original differential equation is of the form

$$\Delta_x^\alpha u(x, y) = \sum_1^\beta a_j(x) \frac{\partial^j}{\partial y^j} u(x, y) \tag{3.70}$$

(y is a scalar, α, β are integers.)

The system (3.68) in the case under consideration will take the form

$$\tilde{v}_k(x^0, \xi) = \int_{D_0} G(x, x^0) v_0(x, \xi) a_k(x)\, dx,$$

$$v_0(x^0, \xi) = \int_{D_0} G(x, x^0) \psi_0(x, \xi)\, dx,$$

$$\tilde{v}_k(x^0, \xi) = -\sum_1^{k-1} \int G(x, x^0) v_{k-j}(x, \xi) a_j(x)\, dx + v_k(x^0, \xi)$$

$$- \int_{D_0} G(x, x^0) a_k(x) v_0(x, \xi)\, dx - \int G(x, x^0) \cdot \psi_k(x, \xi)\, dx, \tag{3.71}$$

$$G(x, x^0) = \begin{cases} \gamma_{\alpha n} |x - x^0|^{2\alpha - n} \ln |x - x^0|, & \text{for } n \text{ even, } 2\alpha - n > 0; \\ \\ \gamma_{\alpha n} |x - x^0|^{2\alpha - n} & \text{in the remaining case.} \end{cases}$$

From (3.71) it follows that for uniqueness of the solution of the inverse problem under consideration it is sufficient that there be a unique solution of the integral equation of the first kind with respect to the function $a(x)$

$$\int_{D_0} G(x, x^0) \cdot v_0(x, \xi) a(x) \, dx = \tilde{v}(x^0, \xi). \tag{3.72}$$

We multiply both sides of (3.72) by an arbitrary function $\varrho(x^0)$ and integrate over the domain D_1.

$$\int_{D_1} \tilde{v}(x^0, \xi) \varrho(x^0) \, dx^0 = \int_{D_0} W(x) v_0(x, \xi) a(x) \, dx,$$

$$W(x) = \int_{D_1} G(x, x^0) \varrho(x^0) \, dx^0. \tag{3.73}$$

The function $W(x)$ in (3.73) is a potential with density $\varrho(x_0)$ distributed in the domain D_1 whereas the function $v_0(x, \xi)$ is a potential with the density $\psi_0(x, \xi)$ distributed in the domain D_2. Hence, for uniqueness of a solution of (3.72) it is sufficient that the linear hull of products of these potentials should be dense in the set of functions defined in D_0.

Let, for the moment, the set of functions $\psi_0(x, \xi)$ be such that the set of potentials $v_0(x, \xi)$ is dense in the set of all harmonic functions regular in some extension of the domain $D_0 - D_{0n}$ *.

From potential theory it is known that the potentials $W(x)$ lie densely in the set of all harmonic functions regular in D_{0n}. It can be easily seen that in this case the linear hull of defined products of potentials is dense in the set of continuous functions defined in D_0. In fact, we consider the harmonic polynomials of order v which are normalized and homogeneous with respect to the coordinates of the points $x - x^1$ where x^1 is an arbitrary point of D_0. It is well-known that the sum of squares of all these polynomials is equal to $|x - x^1|^2$ whereas the function $\{1 - |x - x^1|^2/R^2\}^N$ approximates $\delta(x - x^1)$ in the sphere $|x - x^1| \leq R$. Thus, we have proved the following uniqueness theorem for the inverse problem for equation (3.70).

Theorem. If a family of solutions $u(x, y, \xi)$ of (3.70) is such that the set of potentials $v_0(x, \xi)$ is dense in the set of all harmonic functions regular in the domain D_{0n} (some extension of the domain D_0), then the solution of the inverse problem for (3.70) is unique in the class of continuous functions $a_k(x)$ ($k = 1, \ldots, \beta$).

---

* This condition is fulfilled if, for instance

$$\psi_0(x, \xi) = u(x, 0, \xi) = \delta(x - \xi)$$

where the parameter ξ runs over the domain D_2.

Bibliography

1. HADAMARD: Sur les problems aux derivées partielles et leur significations physiques. Bul. Univ. Princeton, **13**, 1902.
2. —, Le problème de Cauchy, Paris: Hermann 1932.
3. CARLEMAN, T.: Les fonctions quasi analytiques. Paris 1926.
4. —, Sur les systémes linéaires aux derivées partielles du premier ordre a deux variables. CR. Paris, 197,7 (16,8) 1933.
5. GOLUZIN, G. M., and V. I. KRYLOV: Generalization of the Carleman Formulas (Russian). Mat. Sbornik **40**, 1933.
6. NOVIKOV, P. S.: On the Uniqueness for the Inverse Problem of Potential Theory (Russian). Dokl. Acad. Nauk SSSR **19**, 1938.
7. SRETENSKII, L. N.: On an Inverse Problem of Potential Theory (Russian). Izv. Akad. Nauk. SSSR, Ser. Math. **2**, 1938.
8. SRETENSKII, L. N.: Theory of Newtonian Potential (Russian). Gostekhizdat 1946.
9. —, The Uniqueness of the Determination of the Shape of an Attracting Body in Terms of the Values of the External Potential (Russian). Dokl. Akad. Nauk. SSSR **99**, 1954.
10. MALKIN, I. G.: Determination of the Thickness of a Uniform Attracting Layer (Russian). Trudy Inst. Steklov **2**, 4. Publ. Akad. Nauk. SSSR 1932.
11. ZAMOREV: Investigation of the Two dimensional Inverse Problem of Potential Theory (Russian). Izvest. Akad. Nauk. SSSR Ser. Geogr. and Geophys. **3**, 1939 and Nos. **4** and **5**, 1941.
12. BATEMAN: Some Integral Equation of the Potential Theory. J. Appl. Phys. **17**, 1946.
13. VENING MEINESZ, F. A.: Gravity Expeditions at Sea. Delft: Netherlands Commission 1948. Gravimetricheskie Nablyudeniya Na More. Moscow: Geodezizdat, 1940.
14. ANDREEV, B. A.: Calculation of Spacial Distributions (Russian). Izvest. Akad. Nauk SSSR Ser. Geog. and Geophys. **1**, 1947; **3**, 1949; **2**, 1952; **1**, 1954.
15. RAPPOPORT, I. M.: On the Planar Inverse Problem of Potential Theory (Russian). Dokl. Akad. Nauk. SSSR, **28**, 1940.
16. —, On the Stability of the Inverse Problem of Potential (Russian). Dokl. Akad. Nauk. SSSR, **31**, 1941.
17. TYKHONOV, A. N.: On the Stability of Inverse Problems (Russian), Dokl. Akad. Nauk. SSSR, **39**, 1944.
18. —, On the Regularization of Improperly Posed Problems (Russian). Dokl. Akad. Nauk. SSSR **153**, 1 (1963); English translation, Soviet Math. **4**, 1624 (1963).
19. —, Stable Methods for the Summation of Fourier Series (Russian). Dokl. Akad. Nauk. SSSR **156**, 2 (1964); English tran. Soviet. Math. **5**, 641 (1964).
20. —, and C. B. BLASKO: On Approximate Solutions of Fredholm Integral Equations of First Type (Russian). J. Comp. Math. and Math. Phys. **4**, 3 (1964).
21. —, Incorrectly Posed Problems and the Method of Regularization (Russian). Dokl. Akad. Nauk. SSSR **151**, 3 (1963); English trans. Soviet Math. **4** 1035 (1963).

22. TYKHONOV, A. N.: Solution of Nonlinear Integral Equations of the First Kind (Russian). Dokl. Akad. Nauk. SSSR **156**, 6 (1964); English trans. Soviet Math. **5**, 835 (1964).
23. —, Nonlinear Equations of First Kind (Russian). Dokl. Akad. Nauk. SSSR **161**, 1965. English trans. Soviet Math. **6**, 559 (1965).
24. —, Methods for Regularization of Optimal Control Problems (Russian). Dokl. Akad. Nauk. SSSR **162**, 4 (1965) English trans. Soviet Math. **6** (1965).
25. —, Improperly Posed Problems of Linear Algebra and a Stable Method for their Solution (Russian). Dokl. Akad. Nauk. SSSR **163**, 3 (1965); English trans. Soviet Math. **6** (1965).
26. LAVRENTIEV, M. M.: On the Question of Improving the Accuracy of a Solution of a System of Linear Equations (Russian). Dokl. Akad. Nauk. SSSR **92**, 1953.
27. —, On the Accuracy of a Solution of a System of Linear Equations (Russian). Math. Sbor. **34**, 76 (1954).
28. —, On the Cauchy Problem for the Laplace Equation (Russian). Dokl. Akad. Nauk. SSSR, **102**, 1952.
29. On the Cauchy Problem for the Laplace Equation (Russian). Izvest. Akad. Nauk. SSSR ser. Math. **20**, 1956.
    —, Proceedings of the Third All-Union Mathematical Congress Vol. 2 1956.
30. —, Qualitative Estimates for Interior Theorems of Uniqueness (Russian). Dokl. Akad. Nauk. SSSR, **110**, 1956.
31. —, On a Question Concerning the Inverse Problem of Potential Theory (Russian). Dokl. Akad. Nauk. SSSR, **106**, 1956.
32. —, On the Cauchy Problem for Linear Elliptic Equations, (Russian). Dokl. Akad. Nauk. SSSR, **112**, 1957.
33. —, Uniqueness and Stability of Analytic Functions. Helsinki: Suomalainen tiedekatemia 1958.
34. —, On Integral Equations of First Kind (Russian). Dokl. Akad. Nauk. SSSR, **127**, 1959.
35. —, On Some Improperly Posed Problems of Mathematical Physics (Russian). 1962.
36. —, On a Class of Nonlinear Integral Equations (Russian). Siberian Math. J. IV, 4, (1963).
37. —, On a Class of Nonlinear Integral Equations. Material for the Joint Soviet-American Symp. on Partial Differential Equations (Novosibirsk 1963).
38. —, On an Inverse Problem for the Wave Equation (Russian). Dokl. Akad. Nauk. SSSR, **157**, 3 (1964); English trans. Soviet Math. **5** (1964).
39. —, On a Class of Inverse Problems for Differential Equations (Russian). Dokl. Akad. Nauk. SSSR, **160**, 1 (1965); English trans. Soviet Math. **6** (1965).
40. IVANOV, V. K., and A. A. CHUDINOVA: On a Method of Determining Harmonic Moments of Distributed Mass (Russian). Izvest. Akad. Nauk SSSR, ser. geophys. **3**, 1965.
41. —, On Linear Improperly Posed Problems (Russian). Dokl. Akad. Nauk. SSSR, **145**, 2 (1962); English trans. Soviet Math. **3** (1962).
42. CHUDINOVA, A. A.: An Inverse Problem of the Potential of a Single Layer for a Body close to a Given One (Russian). Izvest. Vysshikh Uchebnykh Zavedenii Math., **6**, 1965.
43. LAVRENTIEV, M. M., and V. G. VASIL'EV: On the Formulation of Some Improperly Posed Problems of Mathematical Physics (Russian). Siberian Math. J. (in preparation).
44. IVANOV, V. K.: An Inverse Problem of the Theory of Potential of a Body Close to a Given One (Russian). Izvest, Akad. Nauk. SSSR, ser. math. **20**, 6 (1956).

45. IVANOV, V. K.: On Improperly Posed Problems (Russian). Math. Sbornik **61** (103), 2, 1963.
46. LANDIS, E. M.: On Some Properties of Solutions of Elliptic Equations (Russian). Dokl. Akad. Nauk. SSSR, **107**, 4 (1956); Uspekhi Math. Nauk. **11**, 2 (68), 1959.
47. MERGELYAN, S. N.: Harmonic Approximation and Approximate Solutions of the Cauchy Problem for the Laplace Equation (Russian). Uspekhi Math. Nauk **11**, 5 (71) 1956.
48. TODOROV and ZIDOROV: The Uniqueness of the Determination of the Shape of an Attracting Body in Terms of the Values of the External Potential (Russian). Dokl. Akad. Nauk. SSSR **102**, 2 (1955).
49. DOBRUSHIN, R. L.: A General Formulation of the Fundamental Theorem of Shennon of Information Theory (Russian). Uspekhi Math. Nauk. **14**, 6 (1959).
50. JOHN, F.: Differential Equation with Approximate and Improper Data. Lectures, New York University, 1955.
51. —, A Note on Improper Problems. Comm. Pure Appl. Math. **8**, 1955.
52. —, Numerical Solution of the Heat Equation for Preceding Times. Ann. Math. P. Appl. **4**, 40 (1955).
53. PICONE: Sul calcolo delle funzioni di un variabile complessa. New York: Avad. Press (1954).
54. BERTOLINI: Sul problema di Cauchy per I-equazione di Laplace. Ann. Math. P. Appl. **4**, 40 (1956).
55. PUCCI: Studio col metodo di un problema di Cauchy. Ann. Sc. Pisa **3**, 7 (1953).
56. —, Sul problema di Cauchy non ben posti. Rend Ac. Naz. lincei. **8**, 18 (1955).
57. —, Discussione del problema di Cauchy per le equazioni di tipo elliptico. Ann. Math. P. Appl. **XLVIII**, 1959.
58. KREIN, S. G.: On Correcteness Classes for Some Boundary Problems (Russian). Dokl. Akad. Nauk. SSSR **114**, 6, 1957.
59. KREIN, M. G.: On the Inverse Problem for the Inhomogeneous String (Russian). Dokl. Akad. Nauk. SSSR, **82** (1952).
60. NIRENBERG, L.: Uniqueness in Cauchy Problems. Comm. Pure and Appl. Math. **X**, 1 (1957).
61. CALDERON: Uniqueness in Cauchy Problem. Am. I. of Math., **LXXX**, 1 (1958).
62. FOX and PUCCI: The Dirichlet Problem for the Wave Equation. Ann. Math. P. Appl. **XLVIII**, 1959.
63. PUCCI: Acune limitazioni per le soluzioni di equazione paraboliche. Ann. Math. P. Appl. **XLIX**, 1959.
64. SCHVANK and LIUSTIKH: Interpretation of Gravitational Observations .Gostopteknizdat 1947.
65. MALOVITCHKO, A. K.: Methods of Analytic Continuation of the anomalies of the Force of Gravity and their Application to the Problem of Gravitational Prospecting. Gostoptekhizdat 1956.
66. SUDAKOV, V. N., and L. A. KALFIN: A Statistical Approach to Improperly Posed Problems of Mathematical Physics (Russian). Dokl. Akad. Nauk. SSSR, **5**, 1957.

Gesamtherstellung: R. Oldenbourg, München 8

SPRINGER-VERLAG
BERLIN·HEIDELBERG·NEW YORK

Springer Tracts in Natural Philosophy